More Praise

"Journalist Hannah Nordhaus b̲   ̲ical research and resolute ghost hun̲   ̲the restless spirit of her great-great-grandmother Julia Schuster Staab."
—*Boston Globe*

"Beautifully written and self-aware, a memoir that tells a story and searches for broader lessons. . . . Ultimately, *American Ghost* is not just the story of a haunting but also a story that will haunt its readers."
—*Minneapolis Star Tribune*

"Nordhaus takes us on a journey back in time . . . to draw a better picture of who her great-great-grandmother was." —*Washington Post*

"Nordhaus attacks her subject with the same scholarship and lively writing that made her nonfiction debut, *The Beekeeper's Lament*, a beloved bestseller. . . . Fascinating." —*Dallas Morning News*

"The more Nordhaus digs into the history and explores the supernatural dimensions of the story, the more complex and intriguing it becomes. *American Ghost* is a multigenre work that succeeds on a number of levels." —*Denver Post*

"[A] funny, moving, and suspenseful tale." —*The Week*

"[A] chronicle of German-Jewish immigration to the American Southwest, a reckoning of family secrets, and an account of the author's personal ghost hunt." —*Santa Fe New Mexican*

"In this captivating book, award-winning journalist Hannah Nordhaus examines the story of her great-great-grandmother Julia, who is said to haunt a hotel in Santa Fe as a dark-eyed ghost in a long black gown." —*Buzzfeed*

"An incredible story. . . . A haunting tale." —*National Examiner*

"A unique collision of family history, Wild West adventure, and ghost story. . . . Perceptive, witty, and engaging."
—*Publishers Weekly* (starred review)

"Part travelogue, part memoir, part ghost story, part history. . . . Nordhaus offers a deeply compelling personal account of her attempts to better understand her own family. . . . The book's unique blend of genres and its excellent writing make it hard to put down."
—*Booklist* (starred review)

"The author's multifaceted work brings Julia back to life and explores the journey it took to rediscover her narrative. . . . Every aspect of the account is enlightening, well written, and entertaining. This touching and uplifting work is highly recommended and will appeal to a variety of readers." —*Library Journal* (starred review)

"A thoughtful and intriguing chronicle of familial investigation."
—*Kirkus Reviews*

"A fascinating and nuanced account of her ancestral ghost story and her complicated clan." —*BookPage*

"Fascinating and frequently surprising. Ultimately, *American Ghost* is a reflection on how the unresolved questions in our own histories can be even more haunting than ghosts." —*Shelf Awareness*

"Incredibly moving. . . . *American Ghost* is a treat that appeals to both the mystery-craving and truth-loving parts of the brain. Written with heart, sensitivity, and intelligence." —*5280 Magazine*

"Carefully sifted and exquisitely well told. . . . Here is a very different sort of a Western, a deeply feminine story with a strong whiff of the paranormal—Willa Cather meets Stephen King. Don't read this book late at night . . . unless you like feeling your neck hairs stand up on end!"

—Hampton Sides, author of *In the Kingdom of Ice* and *Blood and Thunder*

"*American Ghost* is several books in one: a true-life mystery, a story of the supernatural told by a skeptic, and a juicy narrative of cultivated European immigrants in a dusty Southwestern town—Santa Fe, New Mexico. Hannah Nordhaus approaches the legend of her great-great-grandmother's ghost with the insight of an historian and the energy of an inspired detective. A fine tale well told. I loved every word."

—Anne Hillerman, author of *Spider Woman's Daughter*

"*American Ghost* is at once an engrossing portrait of a forgotten female pioneer and a fascinating meditation on the fine line between history and lore. Hannah Nordhaus has crafted a seamless blend of gripping mystery, moving family confessional, and chilling ghost story. Once I immersed myself in the stranger-than-fiction life of Julia Schuster Staab, I couldn't stop reading."

—Karen Abbott, *New York Times* bestselling author of *Liar, Temptress, Soldier, Spy*

"Tenaciously researched and beautifully written, *American Ghost* gives flesh to a lost story, exhumes a bygone world, and animates the ways in which the past haunts all of us. Hannah Nordhaus has performed a lyrical feat of dead raising."

—Benjamin Wallace, author of *The Billionaire's Vinegar*

American
GHOST

ALSO BY HANNAH NORDHAUS

Nonfiction

The Beekeeper's Lament

American
GHOST

A Family's Haunted Past in the Desert Southwest

Hannah Nordhaus

HARPER ● PERENNIAL

NEW YORK ● LONDON ● TORONTO ● SYDNEY ● NEW DELHI ● AUCKLAND

HARPER ● PERENNIAL

A hardcover edition of this book was published in 2015 by HarperCollins Publishers.

P.S.™ is a trademark of HarperCollins Publishers.

HarperCollins books may be purchased for educational, business, or sales promotional use. For information please e-mail the Special Markets Department at SPsales@harpercollins.com.

FIRST HARPER PERENNIAL EDITION PUBLISHED 2016.

Designed by Leah Carlson-Stanisic

Library of Congress Cataloging-in-Publication Data has been applied for.

ISBN 978-0-06-224920-3 (pbk.)

16 17 18 19 20   OV/RRD   10 9 8 7 6 5 4 3 2 1

To my father, Bob Nordhaus, who gave me this story,
and so much more

If thou hast any sound, or use of voice
Speak to me.

—William Shakespeare, *Hamlet*

❊ CONTENTS ❊

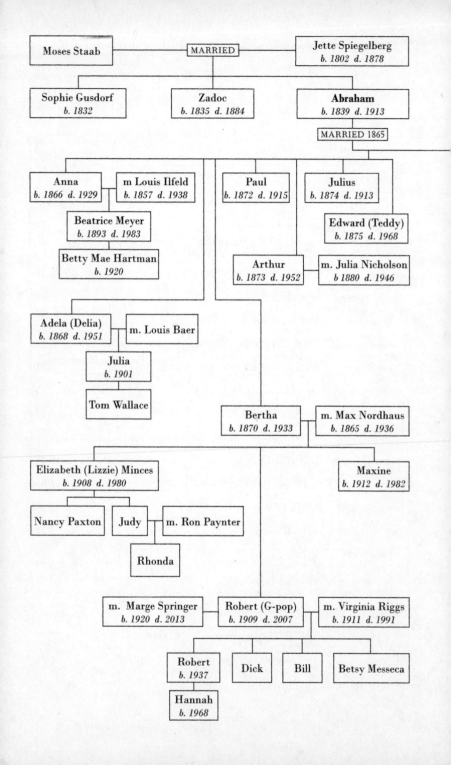

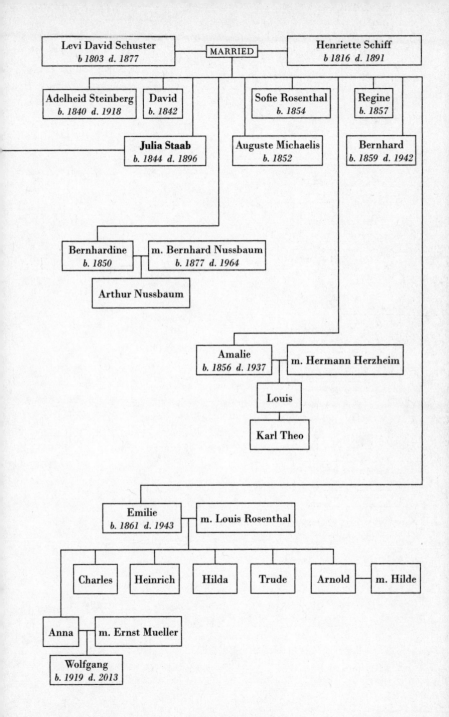

American
GHOST

one

❧ AURA OF SADNESS ❧

The Staab mansion, Santa Fe.

It began late at night, as these stories do. It was in the 1970s, at a hotel in Santa Fe, New Mexico. A janitor was mopping the floor in an empty downstairs room. He looked up from his bucket and saw a dark-eyed woman standing near the fireplace. She wore a long, black gown, in the Victorian style of a hundred years earlier, and her white hair was swept up into a bun.

She was also translucent. Through her vague outline, the janitor could see the wall behind her. She was there and yet she wasn't. She looked into his eyes, and the janitor worried that he might be losing his mind. He looked down and told himself to keep mopping. When he looked up again, she was gone. She had possessed, he told a local newspaper, an "aura of sadness."

The janitor's account was the first the world would hear of the shadow woman in the hotel, but it would not be the last. Some days later, a security guard saw the same woman wandering a hallway. He took off running. A receptionist later saw her relaxing in an armchair in a downstairs sitting room: there, then gone.

Strange things began to happen in the hotel. Gas fireplaces turned off and on repeatedly, though nobody was flipping the switch. Chandeliers swayed and revolved. Vases of flowers moved to new locations. Glasses tumbled from shelves in the bar. A waitress not known for her clumsiness began dropping trays, and explained that she felt as if someone were pushing them from underneath. Guests heard dancing footsteps on the third story, where the ballroom once had been—though the third floor had burned years earlier. A woman's voice, distant and foreign-sounding, called the switchboard over and over. *"Hallo?"* *"Hallo?"* *"Hallo?"*

The woman was seen all over the hotel—in the oldest part of the building, the newer annex, and even the modern adobe casitas that dotted the gardens. One room, however, was a particular locus of activity. It was a suite on the second floor with a canopy bed and four arched windows that looked out to the eastern mountains. Guests who stayed there reported alarming events: blankets ripped off in the middle of the night, room temperatures plummeting, dancing balls of light, disembodied breathing. Doors slammed; toilets flushed; the bathtub filled; belongings were scattered; hair was tugged. People sensed a "presence." Some also saw a woman—the same wan, sad woman—in the vanity mirror, brushing her disheveled white hair.

The hotel was called La Posada—"place of rest"—and it had once been a grand Santa Fe home. The room with the canopy bed had belonged to the wife of the home's original owner. The employees of La Posada were certain that this presence was the spirit of that woman, and that she was not at rest, not at all. Her name was Julia Schuster Staab. Her life spanned the second half of the nineteenth century, and her death came too soon. She is Santa Fe's most famous ghost. She is also my great-great-grandmother.

◇◇

Julia Staab occupies a distant point on my father's family tree—my paternal grandfather's maternal grandmother. She came from Germany with her husband, Abraham, a Jewish dry goods merchant who made his living selling his wares on the Santa Fe Trail. Julia was the young bride he brought over from Germany after he made his fortune. It is said that he built his house—which became the hotel after it passed out of our family—for her. It was a graceful French Second Empire–style home on Palace Avenue, a few blocks off Santa Fe's main plaza—a brick structure in a city of mud and straw, with a green mansard roof ridged with elaborate ornamental ironwork. Julia, too, was formal and elegant, imported from Europe and equally out of place in the rough West: perhaps the house was Abraham's way of making her feel more at home.

I visited La Posada from time to time as a child. My father had moved as a young man from New Mexico to Washington, DC, where I spent my childhood. But we would return each August, when the afternoon drama of mountain thunderstorms painted the desert grasses a gentle green. My parents, my brother, and I would roam through downtown Santa Fe, testing out cowboy hats in the novelty shops that lined the Plaza in the 1970s before the city became fashionable. The hotel was just a few blocks from Santa Fe's ancient heart; we would stop in and look around. The old brick veneer had been covered by stucco to match the rest of the city, and the lush gardens were now pocked with guest casitas. But the interior was little changed from how it must have looked in Julia Staab's time—dark against the brash high-desert sun, cool, filled with Victorian flourishes that still impressed: molded ceilings, ornate brasswork, gleaming mahogany.

I knew, as a child, that Julia had lived in that house, and that the Staabs had once been a prominent family in Santa Fe. They were historical figures in the city, who had brought money and European manners to a place that lacked both. Abraham hosted lavish soirees in the house; Julia played the piano beautifully and spoke many languages. The family had been written up in regional history books; there was even a street named after them in downtown Santa Fe. The mansion was a local landmark. I knew from my father, who was proud of his family's long history in New Mexico if not terribly interested in the details, that Julia and Abraham had raised seven children in the house. His grandmother, my great-grandmother Bertha, was the third.

Beyond that, I knew little of Julia's life. She had lived and died long before I was born, long before even my father and grandfather were born. Although I knew that she had come from Germany, and that her husband had arrived in Santa Fe after New Mexico became a US territory in 1850, I didn't know exactly when or from where. I knew neither how old Julia was when she came to America, nor when she had her children, nor when she died. I knew nothing about her personality and temperament. And I didn't particularly care—I was a child; hers was the story of someone old and dead.

It was around the time I was ten or eleven years old—in the late 1970s—that Julia stopped being quite so dead. A cousin who lived in Santa Fe began hearing stories from neighbors about the shadow woman who popped up in the La Posada sitting rooms and pestered guests in Julia's former bedroom. And accompanying those tales were intimations of a more dramatic nature about Julia's life. Our New Mexico relatives reported hearing rumors around Santa Fe about how Julia had been sickly and deeply depressed, an unhappy bride in a mail-order marriage, shipped against her will or against her grain from the civilized land of her European childhood to the sun-baked hinterlands of the New World.

Julia had suffered in the house, according to the gossip that floated, ghostlike, from Santa Fe bar to restaurant to gallery to stuccoed home. She had lost a child in the room above the La Posada bar and had shut herself in her chamber for weeks. When she emerged, her once black hair had turned completely white. The child was her undoing, the rumors said: after the loss, she became a shut-in. She took to her room and stayed hidden away until she died. Nor was her death peaceful: she killed herself, the stories said—hanged from a chandelier, or overdosed on laudanum. Perhaps she was murdered.

If she died at the hands of another, it was likely her husband, Abraham, who had killed her—and whose persona also underwent a revision. My family had always thought Abraham Staab to be an upstanding immigrant businessman and pioneer. But in the books and articles my aunts and cousins copied and mailed to us in Washington, he became an altogether less appealing character. Now, he kept mistresses and engaged in shadowy business transactions. He frequented gambling halls and bordellos. He was ruthless, "the Al Capone of the territory of New Mexico," explained one article, and he treated his family no better than the victims of his shady deals. Julia was inventory to him, like so much else he bought and sold and owned. He imprisoned her in her room and chained her to the radiator.

My family had once remembered Julia as a cultivated frontier lady married to a daring, civic-minded local millionaire. But now, as my adolescence approached, she became a tortured Victorian wraith imprisoned by a heartless scoundrel. Her story had been remodeled, like the hotel in which she lived. Her life now conformed to the conventions

of a ghost story: the sad woman in a high-necked dress, roaming the
creaking passageways of a Victorian mansion.

◇◇

Of course I found these long-dead ancestors much more interesting
now that there were ghosts and crooks and suffering involved. I had
become a teenager; it made perfect sense that my family—parents and
on up the line—were tyrants and criminals. Julia's legend appealed to
my adolescent sense of melodrama. This was not because I believed
in ghosts. I didn't; I came from a family of rationalists. No, I gravi-
tated to her story simply because it was such a good one. I could now
claim my own piece of the past: a mail-order German bride dragged
west, married badly, driven insane, and trapped forever as a ghost in
her unhappy ending. I loved to tell her tale to my East Coast friends as
evidence of my deep Western roots.

Indeed, New Mexico had always felt like home to me, even though
I had never visited for longer than a few weeks. So at the first opportunity after college, I moved there, into a dust-sloughed adobe in the Sangre de Cristo Mountains, where I read book after book about Western
history and frontier women. I chopped kindling and tended a garden
for the first time in my life, and I began to think of myself as something
of a frontierswoman, too. I reveled in the chamisa and the sunflowers
that climbed the mountains in late summer, when I would sit in my
landlady's meditation garden and watch the tarantulas migrate. It was
a quiet life—too quiet, after a time, for a twenty-two-year-old with
ambitions in the larger world. After a couple of years, I moved to New
York City, where I planned to become a writer and tell other people's
stories. I wore small skirts and big platform shoes and got a job at a
weekly newspaper. I published my very first article a few months later.
It was about Julia.

"I am descended," I wrote, "from a long line of bitter women."

The story told of Julia's ghost and of her suffering at the hands of men. It was light on historical detail, and heavy on self-dramatization and feminist surmise. In the article, I imagined Julia the same age as I was, confronted with the limits of a woman's opportunity in the world. I was struggling in New York and at the paper. It was run by men and full of big egos and go-getters; I felt terribly sorry for myself—and also for Julia. I was certain that she had been disappointed by her choices in life. I was certain that she had been a victim, and that this victimhood lay behind her ghost story. She was a restless, unsatisfied soul, like me—a specter of my twentysomething angst. I wrote in the article that I feared that I exemplified yet another "generation of disappointment," years upon years of women who never got the chances they deserved. After I published the story I decided that the chances I wanted probably lay elsewhere.

I left New York and landed in Colorado, where I felt more at home—and put all thought of Julia to rest.

◇◇

Nearly twenty years later, I was visiting the summer house my great-grandfather had built in the mountains northeast of Santa Fe. One afternoon when I was looking for something to read I found, in a leaded-glass bookshelf, a photocopied booklet—a family history. My great-aunt Lizzie had written it in 1980, shortly before her death, and had made copies for her children and nieces and nephews.

I pulled the book from the shelf, whisked off the dust, and sat to read it on the screened-in porch that faced Hermit's Peak, a lonely wooded crest on the eastern flank of the Sangre de Cristo Mountains, its granite face lurching sharply from the plains. Those same stony cliffs had greeted Julia and Abraham Staab on their journey from Germany to New Mexico. Lizzie was Julia's granddaughter, but Julia had died long before Lizzie was born, and Lizzie discussed her only fleetingly in the manuscript. "Grandmother was an invalid most of her life due to difficulties encountered during her many pregnancies," she wrote. She briefly mentioned that Julia was reputed to haunt the hotel, and that Julia's bedframe could still be found in her old room.

About Julia's husband and the seven children, Lizzie had much more to say. There, Xeroxed and spiral-bound, was a tale of a family ecosystem deeply out of balance—forbidden love, inheritance and disinheritance, anger and madness. There were drug addictions, lawsuits, brother against brother, madhouses, penury, and suicides. There were fatal wounds to the "bosom." These were Julia's children; their story branched from hers. And it was clear to me, from Lizzie's book, that the family was haunted well before Julia became a ghost. I wondered what had gone so wrong.

When I was a younger woman, I imagined Julia to be young as well, struggling as a new bride, frustrated and angry, beset by men. Now that I was older, Julia seemed to be aging alongside me. So many women's stories trail off, in the books and the movies, with the happy ending of finding a mate. But I was married now, and I understood that the wedding is only the beginning. I had children of my own, and I knew the terror of having so much to lose. I knew, too, the daily sturm und drang of raising children. And as I entered middle age, I had come to understand the dread of decline. I would wake up in the night sometimes, gasping, and lie there, terrified of all the loss that lay ahead—the people who would leave me behind, and how as we age we leave ourselves behind, too, and wake up as somebody we don't always recognize.

Motherhood rarely allows for solitude, yet it begets its own kind of isolation: from one's past, from one's youth, from the women we once thought we were and would become. I had, in the years since I'd last written about Julia, gone from heroine to auxiliary, from Fräulein to Frau. I wondered if this was how Julia had felt. She lived in a time before granite-countered subdivisions colonized the empty land, when husbands were masters and women and children were property. But no matter the era, certain inevitabilities remain—we move unremittingly through life from youth to death. I understood this now. Perhaps Julia did, too.

Julia was first an ancestor for me, sepia-toned and indifferent; then a roaming spirit, titillating and abstract; then a symbol of women oppressed. But reading Lizzie's book on the porch swing, facing out to the same mountains that Julia saw as she first approached her married home, I realized that she was also a woman—a specific, particular woman, who courted and married and emigrated and raised children in a rough and unfamiliar world, and then began to grow old far from home and family. Her life was real, and it had traversed the same hurdles and milestones from girlhood to marriage to motherhood to middle age that I was now passing in my own journey. And as I read about Julia's troubled family, I realized that I wanted to know so much more: How had Julia come to New Mexico? How had she found life there? Had she loved Abraham? Had he loved her? Was he the tyrant that the ghost stories made him out to be? Was she ill? Depressed? Insane? Did she kill herself? Was she murdered?

I wondered about the world that she and Abraham had left behind in Germany—why they had left it, and whether their lives were better in this unforgiving land, a place that nonetheless seemed willing to forgive the fact that they were Jews. I wondered what had happened to those who stayed behind. They were Jews in Germany, after all—another story that haunts; whole families erased. In her book, Lizzie described a congenital fragility in Julia's offspring, a tendency to nervousness. I wanted to know how the family had suffered, and why, and whether my ancestors' afflictions might be prone to seeping down through genes and generations.

I didn't think I believed in Julia's ghost, but she was nonetheless starting to haunt me.

◇◇

This, then, became my plan. I would set out on a ghost hunt—a metaphorical one, and a literal one, too. I would come to know the world in which Julia lived and died. I would disentangle the life from the legend, the flesh-and-blood woman from the ghost, the history from the surmise, the facts from the fictions. I would learn about the world Julia inherited and the world she created, and about the children she raised and marked and left behind. I would try to rescue her from the prison of other people's reductions. I would make her real.

Julia would be a difficult quarry—I knew that. Although she was, in her ghost story, a "presence," her life story was riddled with absence. She was a nineteenth-century woman, after all—sequestered in the home, invisible then as now. And I knew of no love letters, no missives to the old country, no diaries written in her hand, no admiring biographies. She had died more than a century before; anyone who had known her was long gone. I could get only so close. To reconstruct her world, I'd have to see her through the eyes and lives of others, in the concentric circles that radiated from the small plot-point of earth she had once trod. I would try to trace those circles—relatives and acquaintances who had once known her and whose own marks on history may also have been light—ever farther from Julia's unobtrusive center.

Her husband, Abraham, was closest. Through Abraham's story, I could surely gain a glimpse into Julia's. Their children, too, had left imprints and memories that had trickled down to us. Like an archaeologist, I could burrow into the layers of evidence my relatives had left behind. I could rummage through Julia's world, and hold those long-buried shards up to the light. And perhaps by reassembling the confused fragments, I could make Julia whole. I could, perhaps, retrieve her from the dark place in which she dwelled.

In this age of information—ones and zeros tracking lives into infinity—there are ways to seek the dead that are more accessible now than they were thirty-five years ago, when Aunt Lizzie wrote about Julia, or twenty years ago, when I did. There are the tools of history: newspaper archives, government records databases, immigration rolls, memoirs, and journals. There are the tools of genealogy: websites, chat rooms, DNA services, online reminiscences, distant relatives blowing around the Internet like dandelion spores and probing shared pasts. For me, there were also living relatives—a few—who remembered Julia's children.

There are, too, more subjective methods available for seeking the dead. I am a journalist and a historian. By temperament and training, I believe in a world that can be measured and tested. But perhaps there was some truth to be mined from the gothic tales floating around Santa Fe: the lost baby, the unhappy marriage, the laudanum, the radiator, the suicide, the ghost in the hotel, stories Western and dark. To unearth Julia, I would explore my Victorian ghost in the company of modern ghost hunters: mediums and psychics, tarot card readers and dowsers and intuitives. Perhaps they could help me find the truth I was seeking.

And finally there was the house. There was her room. She had lived and died in that room, and her ghost was said to dwell there. I would have to visit, of course.

It wouldn't be my first time in her room. I spent a few minutes there once many years ago, soon after I wrote the article about Julia's ghost. My cousins and I had been at a nearby restaurant celebrating my grandfather's eighty-fifth birthday, and we ended the evening in the plush Victorian bar on the first floor of Julia's old home. Someone convinced the manager to let us in to see her room. It had high ceilings, dark woodwork, heavy rust-colored drapes, and complicated furniture. We turned out the lights and stationed ourselves in the armchairs and in the rocker and on the four-poster bed, feeling tipsy and silly and daring at the same time. We called for her as we thought one should when beckoning a ghost. "Julia! *Jooooolia!*" We sat still for a minute or two, felt nothing, and went back to the bar.

Now, though, I would visit with more earnest intentions. I wanted to see Julia's room again. I wanted to spend the night, and wait patiently and quietly for her. I wanted to find her.

But I didn't go right away. I didn't for a while—not until I had hunted across the American Southwest and Germany, rifled archives and history books and the Internet, grilled relatives and mediums and ghost hunters, and learned a lesson about living itself that I hadn't known I was seeking. Nearly one hundred and fifty years after Julia Staab followed her husband into an unfamiliar world, I found myself, finally, back in her room. It had the same four arched windows that looked out to the eastern mountains, which blackened into sky as dusk bled into night. I perched on the end of her bed, and wondered what the night would hold for me.

Misha

◇◇

THEY SAY THAT JULIA lives in the afterworld: in the documented one—history, the remembered past—and also the unaccounted one. I had set out to look for her in both of those places, and while I was comfortable with facts and dates and documents, I had no experience in the world of spirits. I was distrustful and embarrassed.

I decided to start my search with a phone psychic, with whom I could commune from the relative safety of my home office. Searching online for psychics in Colorado, I found myself confronted with a choice: I could select my seer from a website called bestpsychicdirectory.com, or alternatively, from a list provided by the American Association of Psychics. The association sounded more authoritative, so I browsed through the headshots of angel readers, animal communicators, medical intuitives, psychic detectives, shamans, clairvoyants, and Rosemary the Celtic Lady ™, who seemed to be all of the above, until I found someone who caught my eye.

Her name was Misha, and I picked her because she did phone consultations and was affordable—and because she was also very pretty. In her headshot she looked slightly edgy, with china-doll skin, dark, straight hair, and a heart-shaped face. The "About" page on her website explained that her passion was "to bring ALL into LIGHT and to help reveal and heal all that is still in the darkness." The site was built against a starry background with a dormant, witchy-looking tree in the foreground and these reassuring words: "I am real. Accurate. And accredited." That is what I wanted. I made an appointment, shelled out fifty dollars via PayPal, and waited by my speakerphone.

Misha called me right on time. She had a sensible voice—a touch girlish. She told me that she channeled her psychic abilities through tarot and soul cards, holding them in her hands and letting them drop one by one onto a table in front of her, where they would convey messages to her from the beyond. She avoided "full-contact medium work," however—speaking directly to the dead. When she was a child, the dead often contacted her, and this had caused her problems, especially in high school, when it was hard enough for her to speak to living people, let alone the dead. But she assured me that her cards were every bit as accurate. She would watch the cards fall, and tell me what they meant.

This wasn't my first visit with a psychic. I had been once before, when I was in my twenties, late at night after leaving a party in downtown Manhattan. We wandered by a storefront shop and decided, on a whim, to go inside. The psychic looked the Gypsy part: dark hair, dangly earrings, a flowing skirt, scarves. She asked me what I wanted to know. I asked her—of course—if I would find a mate. That was my main concern then. She told me that it would happen, but not soon. She was right about that.

But back then I was speaking—rather tipsily and on a lark—to the future. Now it was the past that concerned me—Julia's past, my family's, this hidden world of memory and myth.

I didn't know quite how to begin. I had trained in graduate school as a historian, and in the years that followed, as a journalist. I was accustomed to matter-of-fact phone interviews: here's my question, quick and concise; tell me the answer; we won't waste each other's time. But my queries now involved dead people floating around in a hidden world that I could neither see nor hear nor understand, and in which, until quite recently, I had had very little interest. How does one interview the dead?

Since I had paid my fifty dollars, I plunged ahead. I offered Misha a vague description of Julia and a haunted hotel. As I spoke, Misha let a card drop. "Well," she said, "the first card that has fallen out is an upside-down temperance card, and the main message with that is that she left this earth in a not peaceful manner."

I heard another card slap down. Misha explained that it was the hermit card. "She's hiding out. She's definitely there and it doesn't look like she's going away anytime soon." She dropped another card, then another. Julia was angry, Misha said. "Her dark side is very much present." The cards kept falling, faster than I could formulate questions. Julia still haunted the hotel because she was missing a piece of herself, Misha explained, and she was going to wait around until it came back to her. "It's like she's quote-unquote stuck in the mud, so to speak"—Misha said "so to speak" a lot, and also "whatnot." Julia had felt trapped in her life. She had wanted to escape.

"What other questions do you have?" Misha asked.

I had so many—who she was, how she lived and died, what it felt like to walk in her shoes—but I didn't quite know how to formulate them. Better to be specific: I asked Misha about Julia's marriage. "Did she love her husband?"

A card dropped. "She definitely loved her husband," Misha said. "But there were other people in her life that she loved as well. She believed that she didn't do anything wrong by loving that person and wanted to have her cake and eat it, too."

A lover?

Yes, a lover. But it didn't last, Misha said. And the cards told her that Julia felt betrayed because of this. There was someone, a trusted close companion "who she thought was safe and whatnot," a male, definitely a male, somebody very close, who "didn't approve of her way of being," and who hurt her in some way.

This was Abraham, I assumed—Abraham, who built a trophy home for his trophy wife, and expected her to be demure and self-sacrificing. Had Abraham driven her into the arms of a lover? Had his disapproval destroyed her in some way?

The cards were silent on this question.

Our half hour was running out. "Is there anything else you want to know?" Misha asked.

I consulted my list of questions. I had only a few left. But they were important ones, concerning the sad events at the end of Julia's life.

"Was she insane?" I asked.

Misha paused. The thing about speaking to ghosts, she explained, is that you get only the perspective of the ghost. The dead don't surrender their subjectivity. So if a ghost doesn't think herself insane, she won't tell you otherwise. "For her it's no, of course not."

I asked her to ask Julia what had happened to her baby—the one she was said to have lost late in life, the one that turned her hair suddenly white.

Misha let a card drop, and gasped. "Oh. Oh my gosh, wow, there are very dark messages coming through," she said, falling silent for a moment. "These are very dark images and cards and messages—that this child was not of the light, so to speak."

Misha said she had rarely received such dark words in all her years of reading cards. "I would say that in ten years this is the second time I've had that, where it's basically pure evil that's come across."

I sat at my steel-and-maple desk, alone with Misha's voice, looking at my world of objects: the burgundy couch against the wall, the books sorted and stacked on the shelves, the afternoon light flooding through the west window. On her end of the line,

Misha saw other things—a sad woman and a baby who was evil, or who made the sad woman feel evil, or who had had evil cast upon her.

Through my speakerphone, we contemplated this long-dead woman and her baby. Misha continued. It was like the baby had "psychically killed a piece of Julia," Misha said.

She hesitated a moment before she spoke again. "This baby," she told me, "was seriously of the darkest stuff that there is."

two

❧ A DRESS OF BLACK SATIN ❧

Julia Schuster Staab as a young bride.

J ulia wore black at her wedding—German brides did then. She married on Christmas Day, and a day or two later, she left her home forever.

In a photo of Julia from around the time she married, she wears a black satin dress with a subtle floral pattern on the arms. Lace cascades down her chest, and a large brooch holds a bow around her neck.

Perhaps this is a bridal photo: she is young, barely a woman, and she sits uncomfortably and looks away from the camera. Her expression is somber, the way people were in early photographs—to be photographed was a serious occasion meant to capture one's image for eternity. Her nose is aquiline and a bit severe. A dark coil of thick, twisted hair is piled high on her head, a frothy fringe of curls circling her forehead.

I first saw that photo when I visited the home of my aunt Betsy, in a winding, rock-and-cactus-littered adobe subdivision thrust up against the Sandia Mountains in Albuquerque, the city in which my father and his siblings grew up. Betsy and her French husband had returned to New Mexico after twenty-five years of living in Europe. She had moved to France after she married—the reverse journey that Julia had taken. Betsy found it difficult to live as a stranger in France and equally difficult to return home after living away for so long. Like Julia, she felt she no longer belonged in either place.

Perhaps that was why our family history had become so important to Betsy: she focused instead on where she came from. She was the archivist of her generation—deeply organized, brusque and businesslike, a tenacious seeker and keeper of photos and stories and newspaper clippings, with no-fuss short hair, a Cleopatra profile, and a thunderous voice. The road to Julia, I was quite certain, traveled through Betsy, who took me in with a big hug and set me up with a pile of files in her spare room, a south-facing rectangle with a trundle bed and a desk. "THERE YOU GO!" she boomed, and she whisked off to wipe the counters, drink her coffee, skype with her children abroad. I leafed through a blizzard of paper: maps and photographs—diminutive strangers in old-fashioned clothing—memos, photocopies, faxes.

It was in that pile that I uncovered a family tree of Julia's kin, the Schusters—nine sisters, two brothers—a large Jewish family. The tree consisted of four pages stapled together to accommodate all the marriages and far-flung offspring, full of odd German names—Emilie, Amalie, Regine, Sofie, Bernhard. Julia was third from the left—"Julie," the German spelling, pronounced *Yul-ya*, born in 1844. Betsy had received the tree from a distant relative she had found on the Internet, a great-nephew of Julia's from Germany. With that many siblings, Julia would have had many great-nieces and -nephews, and until this moment I hadn't realized that I cared to know about them. Betsy gave me copies of the emails she had exchanged with these relatives whose names I had never heard before, whose stories I'd never contemplated as part of my own.

From these exchanges, I learned that Julia's family had come from a village named Lügde—*Luuech-da*—which was then a quiet, compact town of some three thousand people on the Emmer River, in the forested hills of Westphalia in northwestern Germany. The floodplain on which Lügde perches—a set of parallel meadows—offered a clearing in the forest, an interlude of light for those floating downriver toward the Weser and the North Sea. The village was named for that light: *Licht*, *luuech-da*, Lügde. It was tucked in the bottom of a valley between two mounded hillsides, and hemmed in by green patchwork fields and conifer forests—untamed, shady places, full of wild orchids and mushrooms.

There was a roundness to the countryside around Lügde, an alluvial smoothness interrupted by stone farmhouses and solitary castles and gentle watercourses. Lügde was green in the summer. Winters were long. The rock doves, common nightingales, and *eisvögel*—kingfishers—would fall silent around Saint Hedwig's Day on the sixteenth of October, and Julia and her neighbors wouldn't hear the birds sing again until Easter. Saint Hedwig's Day happens to honor a German saint who married young, bore seven children, buried a child, and rose after death—as Julia would.

Julia's family, the Schusters, had been in Lügde for many generations. In her time, the villagers had farmed and produced linen, lace, and cigars. Today it's a still-small factory town best known for its *Os-terrad*, a fiery oak "Easter wheel" that is stuffed with hay, set aflame, and rolled down a nearby hillside. The Easter wheel has been flaming downhill for a thousand years or so—in their own blazing moment, the Nazis embraced it as a true Teutonic ritual. The wheel burned before the Jews arrived in medieval Lügde, peddling their wares from their backs, then from wagons, then from stores. It burned each Easter when Julia was a child; it burned in 1865, the year she married.

The wedding likely took place in Lügde. Julia was twenty-one years old, Abraham twenty-six. Her father, Levi, was a wealthy local merchant, and the wedding would have been a festive affair. There was no synagogue in town—none was permitted. Instead, Lügde's Jews worshipped in a rented house with ten rows of seats left and right of a central aisle, and a ceiling decorated with a starry sky. Perhaps Julia and Abraham made their promises there, or perhaps in her father's large half-timbered *Fachwerk* home in the south quarter of the village, under a pitched roof on a cold Monday evening in December. There would have been feasting and dancing well into the night, the men on one side, the women on the other. In the ketubah—the marriage contract—Abraham would have vowed to provide Julia with food, clothing, and children. In those respects, at least, he kept his promises.

◇◇

Here is how Abraham looks in photos from his youth: clean-faced and compact, with a tightly trimmed beard, a light spray of freckles, deep-set eyes, and a stubby nose. He looks alert and attractive; there's a potency to him. This is the man who took Julia away from Lügde.

Abraham must have known Julia before the marriage, because he had grown up in Lügde, too. His father, Moses, was a merchant there as well—though not, like Julia's father, a particularly wealthy one. In a history of Lügde's Jews sent to me by a historian from the area, I found the Schuster name scattered throughout the document; the family's presence in the village dates back at least to the eighteenth century. But I could find no mention of Abraham's father until 1868, when he served on the village's Jewish council. It seemed the Staabs were relative newcomers to Lügde.

Nor did Abraham stay there long. Born in 1839, he departed in 1854 at the age of fifteen—Julia would have been ten, still a girl. He left in a great wave of migration that swept Germany in the middle of the nineteenth century. The emigrants were young men, mostly, running from the general woes of being a German at that moment in the country's history—famine, conscription, political disillusionment—and the more specific insults that came with being a Jewish German at almost any time: laws and taxes and tolls and proscriptions, reminders at every turn that Jews didn't and couldn't belong.

There was no good reason to stay in Germany, so Abraham left. He would have traveled the Weser River down through the cities of Hameln and Bremen and north on to Bremerhaven, where the Weser emptied into the North Sea and the ships sailed to America. The young emigrants from Lügde went by steamboat if they could afford it. If they couldn't, they drifted downstream for two weeks on floating trees strapped together and sold as wood for shipbuilding at the North Sea ports. From Bremerhaven, Abraham sailed to New York. There's no record of his arrival or of how long he stayed there—according to Aunt Lizzie's family history, he found a position at a merchant house in Norfolk, Virginia, where he stayed for two years, learning bookkeeping and the ancient art of buying low and selling higher. Then he joined his older brother Zadoc in Santa Fe.

They went to work for a cousin, Solomon Spiegelberg, who had come to Santa Fe in 1846 as a sutler, a civilian merchant selling supplies to the US Army, traveling with a column of a thousand poorly trained Missouri volunteers when they invaded Mexico and claimed Santa Fe for the United States. Spiegelberg stayed on afterward, bringing his brothers, five of them, as well as his cousins Abraham and Zadoc. The Staab brothers worked for the Spiegelbergs for two years, learning Spanish and traveling the length and breadth of the territory. In 1859, they opened their own dry goods store and began hauling supplies and capital from east to west over the Santa Fe Trail—brides, too, eventually.

This was Abraham's story—and thus Julia's. By the time he married, he was a rags-to-riches success, a man of business and the world, and his public doings provide one of the few windows through which we can peer into Julia's life. The company he founded with his brother Zadoc, Z. Staab & Bro., was a prominent one in Santa Fe. Wherever I searched in newspapers of the territorial era, I found the house of Staab: multiple ads in each issue advertising new shipments from the "States," lists of wares in full columns and colossal typefaces.

The Staabs sold "stuff," anything a Southwest-bound settler could want or imagine. "Hats Boots & Shoes, Hardware, Groceries &c. &c., all of which will be found as well as sorted, carefully selected and compiled, at the lowest rates." Fur, wool, corn, coffee, sugar, butter, lard, "Common whiskey," "splendid whiskey," beer (Schlitz, exclusively), pianos, razors, saddles. They sold castor oil, calico, "fine custom made clothing," the latest ladies' fashions—though the wagon train across the prairie took so long that the fashions might have changed several times before the clothes arrived—linen cambric, mohair, and garden seeds "at Eastern prices." They sold it all, from first one, then two, then four large storefronts on the Santa Fe Plaza.

Within a few years, Z. Staab & Bro. was the largest wholesale company in the Southwest. The brothers booked hundreds of thousands of dollars—1860s dollars—of revenue each year. Their safes overflowed with silver; what wouldn't fit in the safes they kept in empty ax crates in their office. They made loans, dispensed promissory notes, even issued their own currency.

When the Confederate army invaded Santa Fe in 1862, the Staabs and their Spiegelberg cousins sided with the Union. I could find no evidence that Abraham served in the Union army during the Civil War, though people sometimes called him "Colonel." His campaign was commercial, keeping the Union forts stocked with grain and uniforms. A Confederate soldier who took part in the occupation of Santa Fe wrote of "smooth-faced Jews, that are our bitter enemies and will not open their stores or sell on confederate paper," and suggested that "they ought to be run off from town themselves." Perhaps the Jewish merchants sided with the North because of an antipathy to slavery—or perhaps they simply knew how to pick a winner.

The Confederates didn't hold Santa Fe long. In late March 1862, they fought the Union troops to a bloody draw at Glorieta Pass, twenty miles southeast of Santa Fe. Technically, the Confederates won the battle. But while they were fighting, a battalion of Colorado soldiers happened upon the Confederates' lightly guarded supply train. The Union soldiers looted and torched sixty Confederate wagons, blew up ammunition, spiked a cannon, and slaughtered or drove off five hundred horses and mules. Soon after, the Confederates, lacking ammunition, shelter, blankets, and food—and without smooth-faced Jews willing to supply them with more—straggled back to Texas.

After the Civil War's last shot was fired, Abraham decided to become a US citizen. I found this information in the New Mexico state archives, which are hidden in an industrial cul-de-sac off Santa Fe's busiest thoroughfare. I had driven down from Colorado, following a long stretch of highway that paralleled the route of Julia's own voyage to New Mexico. It was an uncannily hot day. The birds and the flowers were confused, the crocuses lured out of the ground and returned to stalk, the tulips soon to follow. It was painful to go inside and install myself in the archives—a stale, windowless place, as many public archives are. But I hoped that those lightless shelves might shed some light on Abraham, who left Lügde an inconsequential teenager and returned triumphant to claim his bride.

I sat down in the collections room, put on the required pair of rubber gloves, and riffled through a box of folders dedicated to the Jewish "S" pioneers in New Mexico: Seligmans, Spiegelbergs, Staabs. The box was packed with material, but I could find only one disappointingly slim folder on the Staabs. It contained some ledgers detailing corn sales to the army, and an envelope addressed from Abraham to a son. The envelope had the words "Valuable Papers" scrawled in the lower left corner, and when I went to look inside it, the outer flap crumbled in my gloved hands. Carefully, I coaxed the two sides of the envelope open and pulled out a list of names and dates and numbers: "Militia Warrants," it said. The list meant nothing to me. I would learn only later how these warrants had come to obsess Abraham, a dream of easy wealth that nearly destroyed him.

What charmed me at the time, however, was Abraham's citizenship declaration, dated July 10, 1865—four months before his marriage to Julia, who waited in Lügde for her new life. Abraham hadn't, if his meager folder in the archives was an indication, kept much by way of documents to memorialize himself. But he had kept this one piece of paper all his life. It was proof of how far he had come from Lügde, proof that he belonged, and he had tended the paper carefully. A hundred and fifty years later, the parchment was still only slightly off-white. His age, when he signed the declaration, was twenty-six: his "stature" was small—five foot two, the declaration stated—his forehead "low," eyes gray, nose "straight," mouth "small," chin and face "round," hair brown, complexion "fair." His signature was deliberate, an almost childish script, with a big looping "S." With that forceful, fastidious flourish, he secured his status as an American citizen, and promptly went back to Germany to look for a bride.

Was the marriage a family arrangement? A love match? A bald transaction? I imagine that he and Julia had known each other casually before he left Germany, when she was a child and he a teenager. I picture their families greeting each other in the cobbled street after services at Lügde's rented shul, the children awkward in their Saturday finery. Or perhaps their fathers did business together, or the children met in the meadows alongside the Emmer on a windless day, tossing stones that shattered the still reflections of the riverside's willows. Or maybe Abraham didn't remember Julia at all, since she was five years younger and one of many sisters. Perhaps he had simply hired a marriage broker to assist him in his bridal interviews—his *Brautschauen*— and Julia had been the right age and possessed the right dowry. Or maybe she had played the piano for him and had learned a word or two of English, and this endeared her to him.

For the Schusters, Abraham was no doubt an easy sell: poor boy goes to New Mexico, makes a success of himself, promises a glorious American future. In whatever way Abraham arranged the marriage, he worked quickly, as I imagine he always did. The Civil War ended in April 1865; he became a citizen in July; he married Julia five months later. She folded away her high-necked dark wedding dress; packed her steamer trunks with stockings, jewelry, and bridal gifts; said good-bye to her parents, her sisters and brothers; said good-bye to everything and everyone she knew; and departed: Lügde to Bremen, Bremen to Liverpool, Liverpool to New York.

◇◇

They sailed on the RMS *Scotia*, a double-masted, double-smokestacked, red-hulled paddle steamer of the Cunard Line. It was a new ship, only four years old, the fastest on the Atlantic at the time—the journey took nine or ten days. This is the very ship (a fictional version of it, anyway) that Captain Nemo's submarine strikes in Jules Verne's 1870 book, *Twenty Thousand Leagues Under the Sea*, leaving a triangular perforation in the hull—but the ship, being so well built, survives the blow. It was considered the Cunard Line's finest steamer. There was no steerage on the real RMS *Scotia*, only first class. Theodore Roosevelt and his family traveled on the *Scotia* to Europe for their first grand tour there, three years after Julia's passage.

The cabins were located on the main deck, nine feet in height. Two bright and spacious plate glass and mahogany-paneled "saloons" provided dining space for three hundred passengers. Guests on Cunard's luxury ships could expect everything one might order at a fine hotel: halibut, oranges, *petit filets de boeuf à la parisienne*, French beans, littleneck clams, ox tongue, boar's head, galantine of game, mince pie, roast potatoes, neapolitan ice cream, champagne jelly. There was a bakery, a butcher, an icehouse. An onboard medical office was available to treat sick passengers. Above decks, a promenade extended from stem to stern. But Julia's was a January passage, and she was new to the rough winter sea. She may not have needed the doctor's attention, but she probably didn't spend much time out of doors.

Of course, she must have gone above decks as the ship approached New York, to catch her first view of the American coastline: farms, forts, forests, telegraph pylons, homes, the lighthouses of Long Island, and finally New York Harbor and the spires of the city. Julia disembarked on January 12, 1866, and paused in the city to arrange her trousseau, buying clothes and furniture for her new home in New Mexico. Word of her arrival reached Santa Fe soon after. On January 20, this item appeared in the *Santa Fe Weekly Gazette*, with their name misspelled but unmistakable:

Married:—Our townsman, Mr. Abraham Stabb, who went to Europe last year, took unto himself a better half. . . . He went to the

Father Land for the purpose of visiting his friends and relatives, but he did more than he expected, he lost his heart and found a help meet for life.

He has our best wishes for a full measure of happiness in his new estate.

Lynne

◇◇

I N A U G U S T 2012, I spent a night with my husband and children at La Posada. We didn't stay in Julia's room; I wasn't yet ready for a night with the dead. We slept instead in a casita set well away from the house, a thick-walled adobe duplex with a stone patio. We saw no ghosts that night; we heard the groan of a water heater and the snores of our congested daughter—that was it. In the morning, we wandered through the reception area and past the entrance to the old house on the way to breakfast. The grand entry vestibule was dark against the brilliant morning sun, and the children ran quickly past it to the restaurant's patio and the gardens beyond. My husband, Brent, and I ate breakfast while the kids played hide-and-seek among the coreopsis and sunrose, climbing a gnarled apricot tree that, a nearby sign informed us, Julia had planted a hundred and thirty years before.

After breakfast, the hotel's marketing director joined me for a cup of coffee and offered to put me in touch with a writer named Lynne, who had developed an interest in Julia's story while visiting the hotel on a junket for travel journalists. Lynne had done quite a bit of genealogical research on my family, the marketing director explained, and I was thrilled to hear that there was somebody else tracing Julia's path.

When I returned home, I emailed Lynne. She sent me an article she had written about Santa Fe and the hotel, with a sidebar that mentioned Julia's arrival on the "ship 'Scotia' " in 1866. I hadn't known until then which ship Julia had taken. How, I asked her, did she find this out?

Thus began my introduction to online genealogy—ships'

logs and passport applications and death records and third and
fourth and seventh cousins, online family trees of unstable con-
figuration and dubious accuracy, as gnarled as the apricot trunk
on which my children had clambered. Lynne was a talented guide
to this labyrinthine world. She located census records that placed
Julia in New Mexico in 1870, 1880, and 1885 with a rotating cast
of children and family: cousins, clerks, servants, a bachelor uncle.
She traced the Staab family tree from Germany to the United
States and located Zadoc's descendants in New York; she even
tracked Dutch relatives from the other side of my family back to
their arrival in New York in the 1660s.

And Lynne also found the *Scotia*'s log, which included a
handwritten list of passengers—Julia among them. "There she
was," Lynne wrote in her email, "married name and all." The
ship's documents were all scanned and posted online; she told
me I could see them myself. Which I did, and sure enough, there
Julia was, her name scrawled in thick cursive. "Julia Staab," age
twenty-one, hailing from Prussia, the German kingdom into
which Lügde had been absorbed in 1807. Listed right above her
was a companion—not Abraham, as I'd expected, but rather
an "Adolph Staab." A name I had never heard. A brother? A
cousin? "I find it interesting," Lynne wrote, "that her husband
left this duty to someone else."

I needed to know who this Adolph was. I went back to my growing sheaf of records on the Staabs—the papers my aunt Betsy had
given me and the documents I'd located later in the archives and
on the Internet. I dug around in the pile until I found yet another
family tree that covered Julia's and Abraham's descendants in the
United States. This one was created by my great-aunt Lizzie, and
at its top sat Julia and a man named Adolph—Julia's husband, pa-

triarch and progenitor of the rest of the family. Abraham's official name was Adolph. Though many nineteenth-century German Jews still addressed each other with Hebrew names in their households, they gave their babies proper German names for use in the world—Karl, Heinrich, Maximilian, and also Adolph, before Hitler rendered the name permanently unfashionable. Thus, Adolph was Abraham's German name. Abraham was his Jewish one.

Still, it was fitting that Julia's husband had been given a name that was loaded, in hindsight, with such malignant associations. Julia's ghost story had started as whispers in the halls of La Posada and the houses of Santa Fe, then migrated to the newspapers and from there to the ghost books and ghost tours and the Internet. And as Julia's tale grew and morphed, Abraham's ill repute did, too.

Abraham had, in the years after his death, been remembered as a leader and a builder. But now people thought him a far less upstanding man. He was a tyrant and a grasper, who treated Julia as a possession—"arm candy," said one website devoted to the ghosts of Santa Fe. He imprisoned and murdered her, explained another, "so he could resume his powerful position and social standing in the community." This was not the Abraham I had learned of as a child. This was an entirely different creature—the kind of predatory male that all modern women fear.

Lynne certainly believed Abraham to have been such a man. She and I hadn't ever met face-to-face, but our emails grew more and more personal. I learned that she was divorced, not at all amicably, and that she had some bitterness toward her ex-husband. I learned that she lived in Arizona, suffered from neck pain, and had three blond granddaughters. I found her picture online as well—she looked trim and adventurous; in her sixties, probably, with sporty blond hair.

As we shared our histories and information, I came to understand that Lynne's engagement with my family's history was a passionate and visceral one. Through these dry genealogy websites—the census records and ships' logs and family trees—she constructed a far fleshier life for the dead than I dared to imagine. I parsed dates and tidbits, turning them over and over like stones, hoping for a small glint of insight. But Lynne looked at the same stones and built castles—gothic constructions with plots and passions, demons and dungeons, moats and parapets and matrons in distress.

Lynne believed, she told me, "on a psychic level," that Abraham was responsible for Julia's decline. This speculation had "no foundation in any document," she readily admitted. But this is what she concluded: Abraham had chained Julia in her room at the end of her life. He may also have chained her in the basement, because a hotel employee told Lynne that Julia had left "clawed marks in the walls."

This was after Julia had become utterly undone—though Abraham was, Lynne believed, responsible for that earlier unraveling as well. Lynne thought that the "pivotal assault" on Julia's sanity was the death of her child—that same dark baby whom Misha had found so disturbing. Abraham, Lynne told me, had arranged the child's death.

Lynne had arrived at this belief after staying at La Posada the year before—in a casita on the grounds, as I had on my recent trip. In the dark hours of the night, she had had an encounter, and she left the hotel "feeling convinced," she wrote in an email, that something terrible had happened to the baby. Not just its death—something worse.

Three men came to Julia, Lynne explained. They told Julia that she would have to kill the baby. If she didn't, they would.

These men were Abraham's surrogates, sent to Julia's room to do his dirty work. He couldn't abide the sight of the baby. He didn't want it to live, Lynne wrote, because, "Abraham did not believe this last child was his!"

It had been seven years since the previous birth. . . . He had built quite an empire and was not about to allow a "bastard" to pass as his. . . . Julia could not kill her own child, so the men came and drowned the child—I believe—and poor Julia came to retreat into madness.

Lynne told this story with such authority; there wasn't any question in her mind. I asked her how she knew about the men drowning the baby.

A dream, she responded. She'd seen it in a dream.

She had been sleeping in her casita at La Posada and had awoken in tears, unable to shake the vivid images. The nightmare had haunted her for days—it still haunted her.

"Julia," she wrote, "died in her bathtub."

three

❧ THE PRAIRIE OCEAN ❧

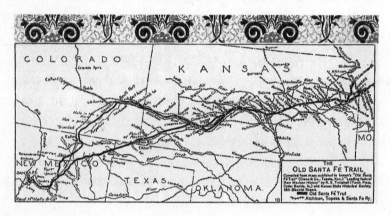

The Santa Fe Trail.

In the real world, things weren't yet so dire for Julia. In the life I could trace through newspapers and archives, Julia was still freshly married and setting out to make a new future in the New World.

Still, the trip to Santa Fe must have involved increasing degrees of shock: the steamboat from Bremen to Liverpool, German-speakers crowding the decks; the luxury liner from Liverpool to New York, with all its strange languages and the ocean, broad and wind-whipped; the trip via train and steamboat from New York to Kansas; the arduous journey across the plains. Julia was twenty-one years old, and had left her family and home for the first time on a journey that was long and irrevocable, with a husband she had yet to decipher. Everything was foreign; everything was new.

When Abraham had first traveled the trail to New Mexico, the trip had been long—two to three months, a laborious, uncomfortable, and uncertain journey following a disjointed thread of commerce and conquest from the jumping-off point in Missouri through the unsettled Indian territories into what had recently been northern Mexico. Julia's trip was faster; the railroad now extended to Fort Leavenworth, Kansas. She and Abraham—Adolph—rode the Burlington, and then the Hannibal and St. Joseph Railroad to its end, and then boarded a steamboat to Kansas City, where the trail began.

It was winter. The plains were colored ochre, the grasses sharp and dead, the view flat to forever across the great speckled plains. In the earliest years, the wagon trains set out from St. Louis, banded together in caravans for safety: wagons and carriages, tents and drivers, hundreds of oxen, mules, and dogs, and very, very few women. Susan Magoffin, an eighteen-year-old bride who traveled the trail in 1846 with her merchant husband, was among the first. She wrote of leaving amid a "cracking of whips, lowing of cattle, braying of mules, whooping and hallowing of men"—as well as other utterances that Magoffin dared not transcribe. The wagons moved slowly west, several abreast on parallel tracks, rolling like cloud shadows across the contours of the land.

Nineteenth-century Germans were voracious consumers of travel literature, especially about the American West, and Julia would certainly have known about the trail. She would have heard about all manner of tribulation and discomfort: about runaway horses, and windstorms and snowstorms and interminable rainstorms, about hailstones "larger than hen's eggs," as Magoffin described them, and swarming mosquitoes that knocked against the carriages like a hard rain, and "troublesome quagmires" that could trap a wagon to its hubs in mud. She might have read of the rattlesnakes: "One hears almost a constant popping of rifles or pistols among the vanguard, to clear the route of these disagreeable occupants," wrote Josiah Gregg, who first traveled the trail in the 1830s and whose books were widely read in Germany.

Julia was lucky compared with these early travelers, and compared with Abraham on his early trips. There was now a stagecoach route from Kansas City. So instead of nine or ten weeks on a wagon train at the plodding pace of oxen, Julia traveled in a Barlow and Sanderson coach, covering the nearly seven hundred miles to Santa Fe in about fifteen days. Not that it was a comfortable journey. The seats were stuffed with hay to keep contusions to a minimum, but it wasn't much help, with the wheels jolting over ruts and pits and stones, the coach moving at the pace of a trotting or cantering horse on a trail not yet a road but only a suggestion of one. Hay lined the floor to warm Julia's feet, and buffalo robes warmed her lap. To keep out the cold air, the side flaps were fastened. Julia rode in the dark.

By the time Julia traveled the trail, there were places to stop for food, water, provisions, and sleep—log-raftered, mud-plastered, dirt-floored huts serving cuisine that insulted the memory of the fine viands of the Cunard Line, or the *Wurst* and *Brötchen* Julia knew from Germany. She ate beans and tortillas, dried buffalo, stew, salt pork; she drank bitter coffee. Kosher was no option here. Julia dined with fellow travelers: cowboys, mostly, toting pistols and straight-bladed daggers. Everyone, cowboys and German brides alike, slept on the floor, or in dirty, bug-bitten lofts reached by rope ladders. Room partitions might be muslin sheets strung from wall to wall. Julia must have been glad to have Abraham, the seasoned plains traveler, by her side.

The winter of 1866 was a season of deep snows on the plains. It was, said old-timers, "the hardest winter they ever experienced." The Arkansas River froze twelve to eighteen inches thick. Twenty wagon trains were halted by blizzards that obscured all sight and passage; stagecoaches were delayed for days. Two trainmen building the railroad that would, in a decade's time, approach the Rockies, froze to death on Bear Creek in Kansas. Natives were also a problem for European settlers. Julia's journey took place during the height of the Indian Wars; the plains were dangerous. After the Civil War, the Indians had stepped up their campaigns against travelers through their territory—emigrants were streaming west and the tribes knew that they would bring only trouble. Caravans were regularly beset by large groups of warriors, and most wagon trains and stagecoaches traveling the trail were now accompanied by federal troops to protect them. The Comanche and the Kiowa were feared above all, and the lore of the trail grew like an attenuated game of telephone, with stories of victims staked to the ground, bellies slit and organs sliced and eaten. Women were warned that capture by the tribes would be far worse than death.

The winter that Julia traveled the trail, six soldiers were scalped four miles from Fort Dodge. A few months later, five nuns—Sisters of Mercy—left for a mission to Santa Fe accompanied by Jean-Baptiste Lamy, Santa Fe's first bishop. About six weeks into the journey, the bishop sensed something. He ordered a corral to be made. All the wagons in the caravan were arranged to form a circle, and the oxen, nuns, teamsters, and other travelers took shelter inside. "None too soon," wrote Sister Blandina Segale, a young nun whom Julia came to know and whose journals were later published in a book, *At the End of the Santa Fe Trail*, "For the Kiowas' death whoop preceded the sling of hundreds of arrows."

The Indians were hidden in the trees across a small stream, shooting their arrows; the priests and teamsters returned fire with guns. It was hot, and the travelers ran out of water. The river was only a few dozen feet away, but the group could not reach it. The next day, cholera broke out. One young sister died. "Whether it was from cholera or fright the victim gave up her soul to God," Sister Blandina wrote. The arrows kept coming. A young teamster also contracted cholera, and began "pitifully calling for his mother." Sister Augustine, an elderly nun, crawled from under one wagon to the next as the arrows zipped and dropped around her. "Sister Augustine reached the dying young man and tried to soothe his last moments as his mother would have done."

A plan was formed: if they couldn't hold out, the men would shoot the four remaining sisters to save them from the unimaginable fate of being captured. But at the next sunset, the Indians withdrew for the night, and when they did, the bishop's men hauled a barrel of whiskey across the stream to a spot near the Indians' camp and quickly retreated to their wagons. The Indians chopped the barrel open with a tomahawk and drank all the whiskey, and while they did, the nuns and their escorts, "like the Arabs, 'Stole silently away.' " The young nun and the boy who had died were buried together farther along the prairie, "strangers in every way."

This was the path Julia traveled.

◇◇

The trail rose too slowly, at a glacial pace, through the furrowed grasslands along the Arkansas River. For Julia, bundled against the raw cold and peering out of the chinks in the window coverings, there was nothing to suggest that her new life would be a comfortable one. The grasses grew shorter, the land flatter, stonier, drier, incised now by winding buffalo paths. A lone cottonwood here, a hidden creek there, a frozen mudhole, a geometric tossing of shrubs, the vacant sky that seemed to crush the earth below it—the land wasn't flat, really, it only appeared that way, swallowing the gullies and rolls in its immensity, size distorted by distance.

Julia had never seen such a land without trees, without water. It was not so much featureless, the novelist Willa Cather wrote sixty years after Julia had traveled the trail, as "crowded with features, all exactly alike." In Lügde, there were no such expansive views—just valleys and gentle peaks, thick woods, occasional breaches of open field. But in a matter of weeks, Julia would now have seen two vast seas: the Atlantic, whipped with winter waves, and this undulating American desert. As Julia's stagecoach crested a small rise, she could, if she peered out of the curtains, see the curvature of the earth.

From Bent's Fort, a ruined earthen trading post at the border between Colorado and Kansas, sand hills began to disrupt the dead plain, the stony knolls stretching progressively higher, the tangled shrubs growing more treelike as the land rose. And then on a day four or five mornings into her trip, Julia would have peeked out of the side flap and seen something new—a silvery strip of azure-footed, white-capped peaks looming like a line of chalky clouds.

The stage passed through Trinidad, a frontier town of dugouts in the foothills of the Rocky Mountains. A few scattered houses ran the length of two blocks, along with a couple of stores, a mud church, a one-room school. The place was a "rendezvous for the outlawed," wrote Sister Blandina of her own journey west shortly after Julia's. Justice there was of the frontier sort: there were regular lynchings of outlaws white and black and Hispanic, dead bodies dangling on display.

Julia climbed her first mountain pass, "the Raton," soon after leaving Trinidad. It was a twenty-mile trail of steep hillocks and loose rocks and scree, and a tooth-jarring ride. At the top, the snow-covered Rockies spread west, the mountains higher than any that a village girl from Lügde could have seen. Descending, Julia and Abraham saw foothills scalloped in snow and dotted with wind-maimed piñon and juniper; they saw the hematite-seeped sandstone glow bloodred in the low winter light.

They passed through Mora, a humble village in a broad valley propped against the foot of Hermit's Peak—the same cliff face that I looked out on as I first read Lizzie's book about the family. Julia saw her first adobe homes in that valley. She ate, perhaps, her first *chili verde*, and rode on, skirting the foothills of the Sangre de Cristos, through Apache Canyon, past barrancas of stone, mountains upon mountains. The largest peak near Lügde was Mount Köterberg, "mutt mountain," nicknamed "Mount Bow-Wow," a broad mound of softened limestone five hundred meters tall and densely wooded all the way to its flattened crest. Now Julia learned sterner labels for the jagged crags around her, named for martyrs and warring Spaniards and Indians: Blood of Christ, Mountain of Thieves, Starvation Peak.

These peaks were rocky, wooded, hazy, volcanic, composed of greens and blues and darker blues, Gothic red-rock mesas and chisel-topped *cerritos* and now and then the lonely inselberg—an "island mountain" of more-resistant rock that rose alone from the flat desert floor—alone, as I always imagined Julia to have been in New Mexico, never quite eroding into place.

four

❊ GOOD-TIME TOWN ❊

Burro Alley, Santa Fe.

Imagine Julia in Lügde, surrounded by family. It was a compact village, walkable from end to end in a matter of minutes, and she was familiar with every building, every cobblestone, every neighbor. The people there knew her, and had since she was a child.

Now imagine her in New Mexico, riding in the stagecoach with her unfamiliar husband, along and among those inhuman peaks, the sky uncompromising, the ground stark and cloaked in snow and the clumped suggestions of the rocks and spiny flora underneath—cactus, greasewood, Spanish bayonet. Most Germans who visited New Mexico in the early days found it unbearable—an ugly, "bleak and sandy high plateau," in the words of one German newspaper.

They found Santa Fe, New Mexico's largest city, to be equally disappointing. Friedrich Adolph Wislizenus, a Thuringian explorer and naturalist, was the first to express his disappointment with Santa Fe in the mother tongue. The city was, he wrote in 1846, merely a collection of "mud-built, dirty houses." Balduin Möllhausen, a Prussian who visited around the time that Abraham arrived in the late 1850s, also noted that Santa Fe held "little appeal." It is true that, later in the nineteenth century, Karl May's fanciful and wildly popular novels about German cowboys and their Indian blood brothers would inspire Germany's infatuation with all things Wild West. But when Julia climbed out of her stagecoach in 1866, Santa Fe was not yet a place that captured German hearts.

It simply didn't seem a *place*. The flat adobe structures were plunked down at random, built of the same dusty red earth that characterized the streets and yards and fields. When Josiah Gregg first advanced toward Santa Fe in 1831, he saw what he thought was a rather unusual collection of brick kilns in the cornfields. A friend corrected him: " 'It is true those are heaps of unburnt bricks, nevertheless they are *houses*—this is the city of Santa Fe.' " It was a colony of mud. "It was possible to be utterly disgusted with it at first sight, second sight, and last sight," wrote the Vermonter R. L. Duffus. "To enjoy it thoroughly one had to have a flair for such things. Literal-minded persons did not, puritanical persons did not."

What Julia thought of the city—whether she had a flair for such things—I don't know. Her new hometown squatted at the western base of the mountains, from which flowed a stream that trickled to nothing in the summertime. It was small compared with St. Louis but substantial compared with the settlements Julia had passed on her way: three or five thousand inhabitants, depending on who was counting. It had been founded by invading Spaniards in 1609 and had been standing for more than 250 years. But it still looked barely there, houses randomly interspersed with cornfields, hay and grit sheeting from the houses' mud exteriors to the ground—"flat and uncouth," Gregg described the layout. Portal-shaded buildings lined a dirt plaza at the center of town. There was the low-slung governor's palace, the jail, the military chapel, and a few shops—an apothecary, a printer, a baker, two tailors, two shoemakers, two blacksmiths, and a handful of Jewish merchant houses, including, of course, Z. Staab & Bro., which was lodged in a mud building like all the others.

The houses that spread out from the Plaza were squat, with thick clay walls and rounded sills. They were dark inside, their interior walls whitewashed with a chalky dust that rubbed off on anyone who leaned against them, the beds rolled up during the day to provide seating for visitors. Into just such a home—mud, with a walled-off courtyard and a carved portal—came Julia, with her silver soup tureens, her fish knives, and her cake services.

Julia's house would have been no different from all the others. The windows would have been small and deep, the door openings covered with coarsely woven horse blankets, the floors lined with buffalo robes. Tallow candles provided light in the evening. There was no proper stove, just a square opening in the corner where the fire was built. A patina of smoke would have stained the ceiling and nearby walls. For "strangers to the country, the customs, and the language" stepping for the first time into such a house, wrote Sister Blandina, "do you wonder that a lonesome feeling as of lingering death came over them?"

The streets of Julia's new city likely held no more comfort. The Plaza was crowded with carts, wagons, teamsters, camp cooks, roustabouts, horses, mules, burros, pigs, and goats. There were cockfights and gunfights. The town was a confusion of commerce, a Babel of languages. There were only a handful of Jews among the Spanish settlers and Pueblo Indians, among the Navajo, Apache, freed slaves, soldiers, veterans, fortune-seekers, herders, cowboys, dry-land farmers, merchants, consumptives, investors, land-grabbers, miners, and shysters who lived there. Church bells pealed riotously at all hours of the day. Letters from family took weeks to arrive; the first telegraph line wouldn't be strung for another three years. Dust coated everything. The streets were piled with garbage, though in that respect—garbage piles and scavenging animals—Santa Fe probably wasn't all that different from Lügde.

What was different to Julia, as she began to venture out and tried to understand this place she must now call home: the language, Spanish, clipped and rat-tat-tat; the food, tortillas, mutton fat, chili con carne made with months-old meat, sun-dried and malodorous. The people were darker than the Jews of Lügde, and their clothing was garish. The men wore brightly striped shoulder blankets—serapes—over cropped jackets and ruffled shirts; they wore high-heeled, silver-spurred riding boots and enormous hats, their tight, silver-studded trousers held up by wrapped silk sashes. The women wore short-waisted shirts with large sleeves, ruffled skirts and rebozo shawls that served as bonnet, apron, veil, and bodice all in one. They didn't, noted an aghast Susan Magoffin, even wear bustles.

There were perhaps fifty Anglo women—white women—in Santa Fe when Julia arrived in 1866. The rest of the women were of Hispanic and Indian descent. They smoked cornhusk cigarettes and danced in the streets, their arms and necks bare, cleavage brimming, faces painted with a white flour paste to protect them from the sun. Their children ran naked. "I am constrained," wrote Susan Magoffin, "to keep my veil drawn closely over my face all the time to protect my blushes." Julia likely avoided Santa Fe's nightly fandangos, packed with dancing women who painted their faces with the bright red juice of a flowering cockscomb—the current fashion.

It was a good-time town. The dance halls had dirt floors, and when the dust got too thick, the music stopped and the serving girls sprinkled down the floor so that the dancing could resume. DON'T SHOOT THE MUSICIANS, begged signs on the walls. Whiskey flowed at all times of day, and on the streets and in the saloons the gambling never stopped. "The governor himself and his lady, the grave magistrate and the priestly dignitary, the gay caballero and the titled señora may all be seen taking their doubloons upon the turn of a card," wrote Josiah Gregg.

Parties often ended in gunfire. There were beatings and lynch mobs and ears bitten off. With the Indian Wars swirling around the city, scalpings were not uncommon, either. The *Santa Fe New Mexican* reported that a man killed his cheating wife with an ax soon after Julia's arrival—but this did not seem at all shocking. There were shootings and pistol duels on the backstreets and in the alleys, even on the Plaza in broad daylight. At breakfast each day—if the newspaper in the nearby town of Las Vegas, New Mexico, is to be believed—citizens greeted each other with a resigned "Well, who was killed last night?" This half-civilized town was situated in a territory of the United States, but it was not yet American.

Reading accounts of Santa Fe in the years after Julia's arrival, I did begin to wonder whether Abraham had been completely honest about his circumstances when arranging for Julia to leave her known world. The home that Abraham readied for Julia—the adobe that she moved into upon arriving in Santa Fe—lay on Burro Alley. The street was only a few blocks long but housed all manner of iniquity: gamblers, drinkers, and whores, obvious and numerous. The city's most notorious casinos and bordellos were all on Julia's street, which also hosted the local market for donkeys and wood. A photo from the 1880s shows a bare dirt road closely lined with rough adobe walls, with cockeyed peeled-bark viga fences, wood-laden burros, and scrofulous men—no women, none outdoors anyway, and not a speck of vegetation in sight. It was no place for any self-respecting bride.

Susan Magoffin, the young diarist and newlywed, lodged on a more respectable street. She stayed only two months in Santa Fe before continuing south into Mexico and finally home to Kentucky, but even in that short time, the city's limited charms wore off. "I am most tired of Santa Fe & do not regret leaving," she wrote. Julia was a slight twenty-two-year-old from a milder world, who knew neither English nor Spanish and nothing at all about the arid, naked place into which she had been imported. She could not comfort herself with the thought that this was temporary.

five

❀ PROMISED LAND ❀

Zadoc Staab (center) with Jewish merchants and Kiowa Indians.

I can find no record of Julia's first years in Santa Fe—no newspaper stories, no archived documents, no memories preserved in her hand. I don't know if she came to love or fear Abraham, or if the marriage came to feel like a home or a prison to her, or if he simply remained a stranger. I don't know if she always found the roughness of her new city to be a trial, or if there were moments when she relished the adventure and came to appreciate the stark beauty of the place.

There were a only a few other Jewish wives when Julia arrived—mostly spouses of the Spiegelbergs and Abraham's brother Zadoc. Some of the Jewish immigrants had wedded local women: one merchant, Solomon Bibo, married into the local Acoma tribe and became its governor. But others imported their mates from elsewhere—Jewish wives who had followed their Jewish husbands who had followed their brothers and cousins to America in a chain of family and village migration.

It must have been a comfort to Julia when her younger brother, Ben Schuster, joined that chain, arriving in Santa Fe six months after her and moving in with the family, to work as a "drummer"—a traveling salesman—for the firm. Among family, perhaps she felt less alone. And it must have been a blow when, just months after Julia arrived, Abraham's brother Zadoc and his wife, Fanny, who had come to Santa Fe as a bride in 1862, moved permanently to New York City, where Zadoc set up as the firm's official East Coast and European buyer. Julia must have mourned the loss of her Santa Fe sister-in-law, who came as close as Julia could get to replacing the sisters she had left behind.

Perhaps in this time she leaned on the other Jewish merchants' wives, who also spoke German and bemoaned the dust and the dryness and tried to re-create some semblance of the world in which they had grown up. They paid calls at each other's mud houses, walking the dusty streets with parasols or traveling by carriage. They took afternoon teas and late suppers.

These women were strangers to each other—all from different villages and cities. But they were German and Jewish in a place where no other women were, and they became their own tribe. The first Yom Kippur service was observed in Santa Fe in 1860, shortly after the first Spiegelberg wife arrived. After the Jewish wives had borne children, a Denver rabbi traveled to New Mexico and "circumcised a large number of children at an advanced age." The hosts, according to an article in the Ohio Jewish weekly *The Israelite*, were impressed with "the scientific manner in which the operation was performed."

This was a ritual that took place across the frontier. Jewish families laid down shallow roots in unwelcoming soil—strangers in a strange land—and hoped that they would thrive.

◇◇

It would not have been unusual for Santa Fe's German Jews to feel as if they were strangers. They had always been outsiders. They had lived in Germany, after all. Though Jews had been in the German states since the time of the Romans—"It is terribly cold and the air is thick with the colossal chill" wrote a tenth-century Jewish merchant who plied the trade routes near Lügde—they weren't at all considered to *be* German. Traders and merchants and moneylenders, they were set apart by their faith and their dress and their mercantile niche and their language—*Judendeutsch*, Yiddish. They were foreigners, an invasive species. It didn't matter how long they had lived in Germany, which during Julia's childhood was not yet a country but rather a constellation of feudal principalities ruled by kings, counts, dukes, bishops, lords, and margraves. Jews were tolerated within that constellation only because of their money. "The town of Beeskow" wrote one Prussian tax commissary to King Frederick William I in 1720, "would like to have a wealthy Jew."

The Jews loaned money to impecunious German sovereigns, bought and sold things that others couldn't or wouldn't, and funded and supplied armies—much as Abraham and the Spiegelbergs would later do in New Mexico. These were the things that Jews *could* do in Germany in the years before Abraham left. What they couldn't do, depending on which principality they called home, was buy houses without special permission; walk on the sidewalks; farm; employ non-Jews; open stores; own stores facing the street; sell meat to Christians; make cheese or beer; ride in a carriage; trade in wool, wood, leather, tobacco, or wine; or practice a guild craft.

In Berlin, kosher Jews were compelled upon their marriage to purchase wild boars bagged on the royal hunting grounds. Later, newlyweds were required to purchase china from the Royal Porcelain Factory—seconds and other pieces that wouldn't sell, unloaded by the factory's manager at above-market prices. It is said that Moses Mendelssohn, the great philosopher of the Jewish Enlightenment and the grandfather of the composer Felix Mendelssohn, acquired twenty life-size porcelain monkeys in this fashion. They were garish, foppish creatures, their paws outstretched as if begging.

In some German principalities, Jews were forced to attend church; in others, they wore yellow "Jew badges"; in yet others, they were required to doff their hats to anyone they met in the street who commanded *"Jud, mach Mores!"*—Jew, show your manners. If a local Jew went bankrupt, the other Jews in the community had to pay his debts. There were separate cemeteries for Jews, of course, and separate gallows.

The rules were different from village to village. Lügde was ruled by the Catholic bishop of Paderborn. The town of Bad Pyrmont, only four kilometers away, was governed by the hereditary prince of Saxony. In Lügde, Jews were tolerated in various trades; in Bad Pyrmont they were forbidden from any business. There were kind sovereigns and brutal ones; good harvests and poor ones; times of health and plagues; moments of quiet acceptance and years of anti-Semitic riots, shop-burnings, and expulsions. "From time to time we enjoyed peace," wrote Glückel of Hameln, a seventeenth-century merchant's wife who hailed from a city twenty kilometers from Lügde, "and again were hunted forth; and so it has been to this day."

The first mention of a Jew in Lügde was in 1598—a man named Salomon, a moneylender in dispute with his debtors. He probably wasn't the first Jew to reside in Lügde, however. Jews were often expelled from German villages, and allowed back in, and, when hard times hit, expelled again. They were blamed for plagues and bad harvests, and accused of poisoning wells, stealing Christian babies (to circumcise them), and using Christian blood for sacramental purposes.

A butcher named "Isaac the Jew"—a Schuster ancestor, perhaps— appeared in the criminal records in late 1651 and for many years after that, accused of such atrocities as slaughtering animals fourteen days before Advent, selling veal to a woman "pretending falsely to be pregnant during the fasting period," beating his wife on Sundays, marrying illegally, and allowing calves to be brought into the city on Palm Sunday. For each of these transgressions, he was issued a fine or thrown in jail (and also issued a fine). Other Jews were docked for similar violations: selling oil, inviting peasants inside their houses, carrying meat outside the city, brawling, mismeasuring, or carrying chalk on Christian holidays.

The German Jews paid taxes, lots of them: they were taxed as Jews when they came to live in a new place, and taxed each year they lived there. When they traveled, they paid an extra "Jew tax" at each town gate they passed. A list of those items subject to customs taxes upon entering the city of Mainz during the eighteenth century included: "Honey, Hops, Wood, Jews, Chalk, Cheese and Charcoal." In Berlin, Jews passed through a gate reserved for livestock and Jews. And paid a tax to do so, of course. They were taxed at births, weddings, and funerals; taxed to open a house of worship, and taxed to keep it open. Those who wouldn't or couldn't pay were "unprotected Jews," and had no right to stay. Isaac the Jew was allowed to remain in Lügde in the mid-1600s provided he paid two talers every year for his "letter of protection" and eighteen groschen for his wife. The eighteenth-century Schusters would have paid for this "protection," too—along with the head tax, and the levy to support the church's sexton and pastor, and the fee required to keep their businesses open, and a hefty "goat's-wage" to keep livestock in town. When Lügde's firefighting equipment fell into disrepair, the two hundred Christians in town supplied the money to pay for four new buckets; the Jews financed fourteen.

Jews in Lügde were relatively lucky, though. They could own their houses there. After the 1740s, they could also own land. But they couldn't farm, unless they could do so with exclusively Jewish labor, which most Jews couldn't. They couldn't practice a craft, like lacemaking or woodworking—this was forbidden by law at first, and later by the guilds, whose charters excluded murderers, thieves, adulterers, blasphemers, and Jews. So they stuck to what was permitted: butchering, peddling goods, trading gold and precious stones, and later, horses. They lent money, but if the amount was more than five talers, they had to make the loan in court under witness, because Jews were known to lie.

By the mid-1700s there were five "protected" Jewish families in Lügde, concentrated in the quarter south of the marketplace. Julia's grandparents were among them. They weren't allowed to attend Christian schools, though they all learned to read, regardless of wealth and status. Until the early nineteenth century the Jews of Lügde had no surnames. Jewish men were named after their places of origin, or their fathers, or both; women were named after their places of origin, or their fathers, or their husbands: Glückel came from the town of Hameln; thus she was known as Glückel of Hameln. Moses was the son of Mendel; he was Moses Mendelssohn. Julia's father was Levi David Schuster: Levi ben (son of) David. His father was David ben Levi.

The practice was confusing, even for the Jews, so when the royal government of Westphalia granted Jews citizenship rights and duties in 1807, it concluded that Jews should be named and counted—all the better to be taxed. Some were named for their villages, and some for their trades: Kramer meant merchant; Kaufmann, too; Staab was a

term for "rod" or "staff" and indicated a person who held some authority; Schuster, Julia's family name, meant shoemaker, though there's no
indication that anyone ever made shoes. A family genealogy suggests
they were so named because their house looked like a shoe.

Lügde's Jewish population peaked at 130 in 1863, at the crest of the
wave of emigration that swept Abraham and Julia—and many Lügde
brothers and cousins—to the New World. By 1871, there were 105
Jews in Lügde. They all left eventually; those who remained would
later, of course, be forced to leave.

◇◇

In New Mexico, the Staabs and Spiegelbergs and Schusters were Jewish by birth but American by choice. If the Jewish community had been
small and insular in Germany, it was even smaller in Santa Fe. The
dry land in which these ambitious merchants settled was a place of remarkable fluidity as Mexican rule gave way to American governance,
a barter economy to capitalism, and community land to fenced plots.
Jews had been dark-skinned in Germany. Now they were "white," at
least when compared with the Indians and mestizo Spanish and free
blacks and Chinese who lived beside them.

In the New Mexico that Julia encountered in 1866, nobody seemed
to care whether she and her husband were Jewish. The newspapers of
the territory—most of them, anyway, and certainly the ones in Santa
Fe—wrote kindly of the local Jews. This was sometimes because they
were advertisers and investors, and sometimes because there simply weren't enough Jews to seem threatening. New Mexico's papers
marked the Jewish holidays ("Many of the best residents are of the Jewish faith and they will thoroughly enjoy the holiday"), and noted with
approval the plans for the Jews to build a synagogue in the booming
town of Las Vegas, New Mexico, which lay seventy miles east of Santa
Fe—"That is right, the more churches the better. Let all the sects be

represented." Editorials in the Albuquerque and Santa Fe papers applauded efforts in Germany "to break down the last barrier separating Jews from Christians," expressed dismay at the periodic anti-Jewish massacres in eastern Europe, and chastised any anti-Semitic screeds they came across—"that the Jews are a charitable race," wrote the *New Mexican*, "is allowed even by those who have the strongest prejudice against them."

Abraham cared that he was a Jew. Marrying within the faith appeared to be important to him: he went back for a Jewish wife, after all, as did Zadoc, and Ben Schuster, and the Spiegelbergs. But there were limits—hard limits—to these Jewish merchants' piety. Their stores remained open on Saturdays, and though most of them closed for the High Holidays, not all did. The Jewish newspaper *Die Deborah* noted a few years after Julia's arrival that only eight Jews showed up for Rosh Hashanah services in the local Germania Hall because the merchants wouldn't let their employees take the time off. "The Almighty Dollar is closer to the Jews of Santa Fe than our holy religion," it lamented. There was, in Santa Fe, no temple, no Hebrew school, no kashruth. Indeed, there seemed to be nothing particularly Jewish about Abraham's life in Santa Fe except the history, and the wife, he brought from Europe. He didn't need Israel; he had already found his promised land.

He was American now. His children would be, as well. I imagine that he wanted the same for his wife.

Joanna

◇◇

ON THE INTERNET, I ran across a blog written by a relative I'd never met before—a third cousin named Robby. "My great-great-grandmother was a profoundly unhappy woman," Robby wrote in a post. A few years before, he had exchanged emails with a writer named Joanna Hershon, who was researching a novel based on the Staabs. It had recently been released.

The book is called *The German Bride*, and it tells the story of Eva Frank, a wealthy German Jewish girl who falls into a relationship with an attractive, if morally unappealing, Gentile painter in Berlin. The affair ends in a horrible accident, and Eva's grief propels her into the arms of a dapper, if morally unappealing, local boy gone west, Abraham Shein, who has made his fortune selling dry goods in Santa Fe and has returned to Germany to find a bride. Seeking to escape her sorrow and lured by promises of adventure and wealth in a place far away, Eva marries him, travels the wagon trail, and arrives in Santa Fe. It turns out, though, that Abraham isn't the great success he has represented himself to be. Nor is he, as Hershon puts it, "the most fiscally conservative man in town." Rather, he lives in an adobe hovel on Burro Alley, "scraping along in squalor amid large insects, peculiar cooking smells, and refuse from chamber pots."

There isn't even a bathtub in the house—Abraham Shein orders one for Eva, the first in Santa Fe—but there is nowhere to put it, so it gathers leaves and rainwater in the back courtyard. Later, Eva and Abraham conceive a child in that tub. In Joanna's book, Abraham isn't a kind man: he doesn't allow Eva to keep kosher, and he has a gambling, boozing, and whoring problem.

He is deeply indebted to the madam across the street, and he measures his character by "the fact that he hadn't ever come close to pawning his wife's jewels."

This is an imagined version of Julia's world, her story transfigured by imagination, supposition, and history—through art. But Joanna's interpretation of Julia's life isn't all that different from those stories handed to me by Misha the psychic and Lynne the genealogist. Abraham was a cad; Julia was his prey. She had hoped for love in America; instead she found treachery.

Was this how it had been between Abraham and Julia? Was there any love between them? Did they share anything besides a bed? In Julia's day, marriage was a contract, arranged to deliver the basics necessary for survival and reproduction; it was not a celebration of passion and compatibility. Did Julia have any right, in her time and place, to expect such things?

I was not, of course, the only member of my family who speculated on these matters. Everyone was intrigued by Julia's ghost story. The older generation joked about it without much conviction; we younger ones gossiped and plotted visits to her room.

We were all haunted, in one way or another, by the notion of Julia marooned in the desert, and many of us found in Julia a muse and a metaphor. My mother, a poet related to Julia not by blood but by marriage, composed a poem some years ago about her famous in-law. "This harsh land with its alien colors, flowers sheathed in spines, sky breeding clouds above the sword-encircled blossom . . ." A third cousin, Kay, wrote a children's book titled *Jews of the Wild West*. In it, she explained that Julia was sad, because a child had died in a third-floor fire.

Now that I was searching, I stumbled across fellow Juliachasers without effort. Some were related to me; many weren't. But each time, I was surprised to learn that others felt as connected to Julia as I did—that they, too, had made Julia's story their own and embroidered it with their own preoccupations, as Joanna had in her novel.

In his blog, my distant cousin Robby remarked on the differences between Joanna's novel and Julia's real life, between the imagined Abraham and the one who lived in history. It was, he wrote, "like that *Star Trek* episode where they go to a parallel universe, and see what life would be like if all the crew members were violent and evil at heart."

We create imaginative worlds, parallel universes that serve our needs—I did, when I was a yearning, frustrated twenty-something looking for villains. Lynne did, too, in imagining Abraham as a monstrous man not unlike her ex-husband. Joanna did in her novel. I don't think it's a coincidence that Julia's ghost story gained traction in the 1980s and '90s, as American feminism began to contest more forcefully the notion of submission in marriage, redefining it as abuse. This was the era of madwomen in the attic and burning beds. And in the contemporary version of Julia's story—her story as we modern women have told it—Julia was a victim. And Abraham was a villain.

I don't know that anyone in my family believed Abraham to be a saint, but I got the sense that Robby felt Abraham might not be as bad as Joanna—and Lynne, and many of the ghost stories—made him out to be. If Abraham was domineering, or consumed in his work, or if he gambled and frequented bordellos or yelled at or ignored his wife, was he a villain, or simply a man of his time and place? Would Julia have thought her husband a monster and a scoundrel, or would this be how she expected husbands to behave? Is it fair to judge as villains these ordinary men of an earlier era, simply because they played by rules we no longer honor?

There is the Abraham who raised a proud American family and helped build an American city, and the Abraham who served only himself; there is the Julia who lived in the world, and the ghost woman who lives on in our minds. One feeds the other, and sometimes they intersect.

six

❧ BOOK OF PRAYER ❧

Julia and Abraham Staab, early in their marriage.

I have a photo of Julia and Abraham taken around the time she first arrived in Santa Fe. Abraham sits on a tassel-trimmed chair wearing a dark suit, while Julia stands behind him, leaning in slightly, her hand on his shoulder. Abraham looks straight at the camera, fearless, with a bare hint of a smile. Julia looks neither at him nor at the camera but somewhere between, perhaps at someone else in the room. She wears a full-skirted satin dress embossed on the shoulders and collar, and her hair is coiled smoothly above her head; her eyes are a tad too close together for classic beauty. Her fingers on Abraham's shoulder are relaxed. It appears to me as if there is some affection there. But in another photo taken around the same time, she sits by herself, wearing an even fuller skirt and a smart white kerchief around her neck. She looks warily—wearily— and frankly at the camera, as if accusing. She looks so very alone.

◇◇

Julia's first daughter, Anna—conceived soon after Julia moved into the adobe house in Santa Fe—was born in November 1866. Anna's name is entered on the blank pages of a book still kept in my family—*Neues Israelitisches Gebetbuch für die Wochentage, Sabbathe und alle Feste zum Gebrauche*—The New Israelite Prayer Book for Weekdays, Sabbath, and All Holidays. The prayer book served as the equivalent of the family Bible for Julia, a repository of important family milestones, recorded in German. It had been published in Berlin in 1864, perhaps given to Julia upon her engagement to Abraham. On the overleaf there are six ruled lines in light pencil, the entries written in cautious fountain-pen calligraphy—Julia's hand, I suspect.

The first line recorded Julia's firstborn: *"Anna Staab, geboren am* [born on] *23 November 1866."* Adela, nicknamed Delia, came next, in 1868. Bertha, my great-grandmother, was born in August 1870—the third girl in a row. This can't have been an entirely welcome development. A succession of daughters, in the Old World or the New, was reason for consternation—where were the male heirs?

But at last, in 1872, came Paul, the first of Abraham's sons. "Born," the *New Mexican* reported on January 15 of that year.

> On Sunday morning the wife of A. Staab, Esq., of this city, was safely delivered of a son. Mother and child, we are gratified to announce, are doing well, and the happy father is doing as well as

could be expected under the circumstances. We extend our congrat-
ulations to the parents, and particularly to the father, for we know
it is just what he most desired. We trust that the child, though born
in this time of violence and revolution, may be a perpetual source of
joy to his parents.

I noticed that while Abraham is mentioned by his first initial, Julia appears only as his "wife"—and also that "what he most desired,"
was clearly a son—daughters weren't sufficient. I also wondered what
"revolution" the paper referred to, and what violence—the usual dance
hall stabbings, or the Indian Wars, or something else? The newspaper
article didn't tell me; I couldn't know. There was one thing, however,
that I knew, that the newspaper—and Abraham and Julia—could not
have known: this son would not be the "perpetual source of joy" the
newspaper anticipated. Paul suffered from severe epilepsy, and he required an attendant his whole life. He was, explained one family tree,
"of unsound mind"; a woman who knew the Staabs as a child described
him in an oral history as having been "retarded." What sadness this
must have brought Julia and Abraham when they came to understand
his infirmity.

There wasn't time to lament, however. The next boys arrived in
breathless succession—a boy every year: Arthur in 1873; Julius in
1874, his name honoring his mother's. The fourth and youngest son,
Edward—Uncle Teddy, my family called him—came in 1875. The boys
were given English names, nothing Hebrew about them, and all of them
shared the same middle initial: "A." I know from an old passport application I found online that Teddy's middle name was Adolph, but I could
never figure out what all the other A's stood for. I wondered if Abraham
had named them all after himself—and I also wondered if he later despaired that none of the boys seemed to take after him in any other way.

If there wasn't perpetual joy, there was surely perpetual motion in
Julia's life in those early years. The newspapers' social reports are quiet

on the subject of Julia in those years when her children were small—but it can't have been quiet in her dirt home on Burro Alley. She had a houseful of young children, one birth and then the next, seven children in the first nine years of her marriage. Each one, I fear, took a small piece of her. By the time Teddy came, the names had outgrown the penciled lines in the prayer book—Teddy's dangled off into the white below, the inked script of his name blacker and heavier, as if the writer of it was now unduly burdened.

Those three girls and four boys weren't Julia's only pregnancies, either. A historian named Floyd Fierman, a rabbi who wrote books about the pioneer Jews in the Southwest, mentioned fifteen pregnancies, total—eight full-term, seven miscarriages. This may have been standard for women of her day, in the age before widespread birth control and modern medicine, and Julia wasn't without help in recovering from her confinements. The family had grown rich and richer yet: the Staab & Co. wagons kept coming along the trail, forty, fifty at a time, bringing the world to Santa Fe and supplying the federal troops who stayed on after the Civil War to fight the Indians at a series of forts in and around the city. The Staabs certainly had enough money now to pay for nannies and cooks, maids and wet nurses.

But the help wasn't enough. Abraham's money wasn't enough. Julia was alone in a small adobe house, raising children among neighbors who spoke English and servants who spoke Spanish. Her German-speaking husband was gone all day, out at the store or visiting customers or wheeling and dealing around the territory. She was far from everyone and everything she knew well.

◇◇

On a raw winter day just before New Year's, on my way to the archives at the New Mexico History Museum, I stopped at La Posada. It was a quick trip, with no time for anything but a brief walk around the lobby. I circled once through the bar—the old family sitting room. It was empty so early in the day, and it smelled vaguely of beer. In the hallway, I admired an intricate brass chandelier, then went to the foot of the stairs, where Julia's ghost is seen so frequently. I rubbed my hand along the mahogany banister. Julia's hands had once grasped that curved wood; her children's had, too, grazing the top as they ran up and down the stairs. I felt the heft of the past, indifferent to my presence.

Then I headed off to the archives. I dodged snow piles pushed up against the curbs of the streets along the Plaza to reach a modern two-story adobe building. I climbed to a second-story archive room and began leafing through the library's folder on the Staab family.

In it, I found a letter from a descendant of one of Julia's younger sisters, Sofie Rosenthal, explaining why Abraham had left ten thousand German marks to her in his will. Until then, my suspicions about Julia's depression during her early years in New Mexico were all based on hearsay and supposition. But the letter explained that Sofie had, during the 1870s, come all the way from Lügde to Santa Fe to help during the time when Julia was bearing child after child. Abraham had sent for her. Childbearing had worn on Julia, and she needed more comfort than hired help could provide; she needed family. So Abraham brought Sofie to help. The letter was the first concrete evidence I found of Julia's distress.

Sofie traveled by train to somewhere near Trinidad, where the railroad then ended, and by stagecoach to Santa Fe. But she "only stayed a few years, as she felt so isolated in Santa Fe," the letter said. Sofie wasn't married yet; there was no reason she couldn't return home to Germany to be among the people she knew and loved. Julia remained in the desert.

seven

❊ BRONCHO MANEUVERS ❊

Sister Blandina Segale.

I n early 1877, Abraham sought additional assistance for Julia. He was still concerned about his wife's condition. He wasn't able to send to Europe this time, so in place of a real sister, he decided to procure a Catholic one. In March 1877, Sister Blandina—the young nun whose diaries recounted her days at the end of the Santa Fe Trail— was asked by her superiors to look after Julia. "A lady, Mrs. Adolph

Staab, and her children are here," Blandina wrote in her diary. "I have been asked to entertain them after school hours. I am perfectly at home with the children, but I have no attraction for entertaining wealthy ladies. However, since it is given me as a duty, I'll do it. Mrs. Staab really needs attention. She is in a depressed condition, and I must cheer her up."

For a few weeks, Blandina spent afternoons with Julia. She must have done an adequate job of improving Julia's mood, because in late April, Abraham asked her to accompany Julia and the children to Germany. Julia planned to travel there to see her family.

But Sister Blandina found even the suggestion an affront. "I believe he thinks money can do anything and he expects me to accept the offer," she wrote. "When he was convinced that I could not go to Europe, he said he would be satisfied if I would accompany them to the city of New York. But I am satisfied to remain in Santa Fe."

Sister Blandina's superiors, however, were not satisfied: at the end of May, she reported a compulsory change of heart. "Sister Augustine tells me that the most Rev. Archbishop Lamy"—the archbishop of Santa Fe—"wishes me to go with the Staab family to the terminus of the railroad." The railroad was now five miles from Trinidad, the scruffy frontier town at the base of Raton Pass. It was a five-day, two-hundred-mile stagecoach ride from Santa Fe, much shorter than the seven hundred miles Julia had traveled on her first journey across the plains, but a difficult trip nonetheless. The plan was for Blandina and another young nun, Sister Augustine, to ride with Julia and two of her children; the others must have stayed behind in Santa Fe, or perhaps traveled ahead of them. The women would be accompanied in another stagecoach by Julia's physician, Dr. Symington, Abraham, and "two gentlemen who are going to Chicago." Men in one coach, women in another.

It was a hazardous time on the trail. Billy the Kid was terrorizing Colorado and northern New Mexico with a gang of criminals, stealing horses, robbing stagecoaches, and raiding settlements, guns ablaze. (This was not the famous Billy the Kid, but a less famous Colorado outlaw who preceded him—also young, also named William, also rampaging in the late 1870s.) "Everyone is concerned about our going," Blandina wrote. "Mr. Staab spoke to Sister and myself about the danger of travel (at the present time) on the Santa Fe Trail, owing to Billy the Kid's gang. He told us that the gang is attacking every mail coach and private conveyance." Abraham wanted to make sure that Blandina was comfortable with the prospect of a run-in with the region's most notorious outlaw. " 'We will have many freight wagons well manned, but if you fear to travel, we shall defer the trip,' " he told her.

Blandina found Abraham's chivalry touching, but she told him that she had "very little fear of Billy's gang." This was not only because of her abiding faith in God, but also because she knew—though she didn't explain this to Abraham—that she could offer the family some protection: she was already acquainted with Billy the Kid.

Only a few months before, in late 1876, she had been teaching at a church school in Trinidad when word came that Billy's gang had been wreaking havoc on the other side of the mountains. One of Billy's thugs had "painted red the town of Cimarron," Blandina wrote, "mounting his stallion and holding two six-shooters aloft while shouting his commands, which everyone obeyed, not knowing when the trigger on either weapon would be lowered." A few days later, the same gunman arrived in Trinidad. Sister Blandina watched him approach from her schoolyard. "The air here is very rarified," she wrote, "and we are all eagle-eyed in this atmosphere."

We stood in our front yard, everyone trying to look indifferent, while Billy's accomplice headed toward us. He was mounted on a spirited stallion of unusually large proportions, and was dressed

as the Toreadores (Bull-Fighters) dress in old Mexico. Cowboy's
sombrero, fantastically trimmed, red velvet knee breeches, green
velvet short coat, long sharp spurs, gold and green saddle cover. A
figure of six feet three, on a beautiful animal, made restless by a
tight bit—you need not wonder, the rider drew attention.

The thug passed on through the town, but a few weeks later a member of the local "Vigilant Club" had come to fetch Sister Blandina. "We have work on hand!" he told her. The same outlaw—Schneider was his name—was again in the vicinity. This time, however, he was doing no parading. He had been shot in the thigh after a quarrel with his partner, a man named Happy Jack. The two former compadres had been "eyeing and following each other for three days, eating at the same table, weapon in right hand, conveying food to their mouth with the left hand." Finally, at dinner one night, there came a lull in which each man thought the other off guard. They fired simultaneously, and both were hit. Happy Jack was shot "through the breast" and killed. Schneider was grievously wounded, nursing his injuries in an unused adobe hut in Trinidad. "He has a very poor chance of living," the man from the Vigilant Club told her.

But Schneider lived for quite some time. Sister Blandina had vowed, as a Sister of Charity, to care for anyone in need—outlaws, despondent Jewish wives—and so she brought him food, water, castile soap, and linens, and she ministered to him for months. The two developed a friendship of a sort. He confessed his sins—they were many, including befriending and then murdering inexperienced travelers on the Santa Fe Trail, scalping an old man who had once shown him mercy, and shooting cows for their hides—but he did not repent. "What will my pals think of me?" Schneider told the Sister when she spoke of absolution. "Me, to show a yellow streak! I would rather go to the burning flames!"

A few weeks later, Schneider's pals, including Billy the Kid, showed up in town, planning to scalp four Trinidad physicians who had refused to extract the bullet from Schneider's thigh. Sister Blandina met with them in Schneider's sickroom. "The leader, Billy, has steel-blue eyes," she wrote, "peach complexion, is young, one would take him to be seventeen—innocent-looking, save for the corners of his eyes, which tell a set purpose, good or bad." She declared the rest of the gang, "all fine looking young men." Billy announced that he wished to repay Sister Blandina for her work caring for Schneider. Blandina was six years younger than Julia, but she trafficked easily in the currency of the gritty frontier. "I answered, 'Yes, there is a favor you can grant me.'"

The favor she asked was that Billy refrain from scalping the Trinidad physicians, and though he wasn't too keen on it, he agreed and stood by his word. "Not only that, Sister," he told her, "but at any time my pals and I can serve you, you will find us ready."

The gang rode off, and a few weeks after that, Schneider, Sister Blandina's "poor desperado," approached "the shores of eternity," she wrote. "He has become more thoughtful, even his tiger eyes are softening." She fetched his mother, who helped care for him, and in early December 1876, he echoed her prayers, "which included an act of contrition"—repentance at long last—and said good-bye.

◇◇

Abraham often saw luck go his way; he couldn't have known how smart a decision he made when he insisted that Sister Blandina travel with the family to Trinidad.

The group—the two Sisters, Abraham, Julia, her two children, the two Chicago-bound travelers, and Dr. Symington—left Santa Fe on the first or second of June in a caravan of well-armed freight wagons. They stayed a night at the Exchange Hotel in Las Vegas, New Mexico, where Sister Blandina received an unsavory proposition "to leave the convent and go out to enjoy some of the pleasures of the world. Do you wonder at my indignation?"

They traveled on through Tiptonville, a "map-forgotten place" with a dozen mud-thatched adobe houses, and then through Agua Dulce (Sweetwater), arriving safely in Trinidad on the sixth of June. "The greater part of the day was given to making things comfortable for Mrs. Staab and children to travel to New York City," Sister Blandina wrote. The next morning, Julia and the children boarded the train.

Abraham and Dr. Symington planned to return to Santa Fe, and Sisters Blandina and Augustine, their duties to the archbishop dispatched, were also free to go home. Seeking a speedier return, Abraham and Dr. Symington asked if the nuns were willing to travel back to Santa Fe with them in a hack—a smaller, faster four-wheeled coach. They hoped to break the record for travel from Trinidad to Santa Fe. There were risks, Dr. Symington explained, because "the Kid" was attacking "coaches or anything of profit that comes in his way."

Sister Blandina again told the men that she had no fear. On June 10, they set off, arriving that evening at the stage station at Sweetwater. "It did not take us long," Sister Blandina wrote, "to see that extraordinary preparations were being made." The stage driver and his passengers were loading and cleaning their revolvers, and everyone expected Billy and his gang to attack that night. Blandina, unfazed, took a long walk in the fields outside the station. The next morning, they left early. About an hour or so after lunch, the coach's black driver yelled back to Abraham, his voice "trembling with suppressed fear," as Sister Blandina described it. "Mas-sah," Sister Blandina wrote. "There am som-un skimming over the plains, coming dis way."

Abraham and Dr. Symington took out their revolvers. The rider came closer. "By this time both gentlemen were feverishly excited," Blandina wrote. How I love the image of Abraham, all five-foot-two Jew of him, with his suit vest and watch fob and German accent, hanging from the coach, steely eyed, revolver at the ready. "I looked at the men," wrote Blandina, "and could not but admire the resolute expression which meant 'To conquer or die!'" But Sister Blandina advised that Abraham and the doctor should instead "remain passive," and put their guns out of sight. "They looked at me as if to say that a woman is incapable of realizing extreme danger. The darkey in his fright spoke again: 'He am very near.'"

Sister Blandina again advised them to put their guns away. Abraham listened to her this time—to a woman!—and the men put their guns down. A "light patter of hoofs" drew near the carriage opening. Sister Blandina looked out, and shifted her bonnet so that the rider could see her, as she suspected he had the night before when she took her walk. "Our eyes met; he raised his large-brimmed hat with a wave and a bow, looked his recognition, fairly flew a distance of about three rods, and then stopped to give us some of his wonderful antics on broncho maneuvers." This was Billy, of course.

Now free of desperadoes, the foursome rushed on to Santa Fe. They arrived on June 12, at a breakneck pace. "The record is broken," Sister Blandina wrote. "We made the fastest trip ever known from Trinidad to Santa Fe": the doctor, the nuns, and the Jew.

◇◇

Julia's return to Santa Fe was less hurried. She moved on to New York and from there to Germany, where she stayed for many months.

I know few specifics about her trip there, only that her children reported that for various periods in the late 1870s and 1880s, Julia retreated to Germany for months or years at a time. She had been in New Mexico for more than a decade now, but still it wasn't home. While in Germany, her children attended proper German schools, where they received a proper German education. Abraham visited for a few months in the summer. Whether this pattern was unusual for wealthy immigrants at the time, I don't know.

I know only that Julia was unhappy in New Mexico, and that Abraham was worried and sought to help her. He enlisted sisters and Sisters to come to her aid, and when they couldn't help, he sent her to Germany to heal—and even this was not enough.

Steve

◇◇

IN A 1994 TELEVISION show, Abraham comes across as rather more dignified than I imagine he did waving his revolver in the carriage with Sister Blandina. The show, *Unsolved Mysteries*, opens with an infrared shot of a green-painted hallway. "No crime has been committed. No one has been hurt. No one has disappeared," says the show's host. "Believe it or not, this team of investigators is looking for . . . a ghost."

In a pink-painted room, a ghost hunter with a densely gelled rockabilly haircut tinkers with some equipment—half a million dollars' worth, the host says: huge computers, unwieldy boxes. This is a "classic haunt," the ghost hunter says. "If they find a ghost," the show's host adds, "she is probably named Julia Staab."

Abraham appears, top-hatted, taller, darker, and fuller-bearded than the one I have come to know from family photos. Julia—elongated and paler, with a nest of dark curls piled on her head—steps out of a carriage in front of the mansion. A uniformed maid welcomes them into their new home, blessing them with a heavy Yiddish "Mazel tov!" Abraham beams at Julia. A local historian tells of elaborate soirees held in the house.

"One would think she was a very happy woman," the show's host says, but "Julia's joy turned to sorrow overnight" at the loss of her baby. There is a scene of Abraham comforting Julia over an empty crib, both of them weeping. But she cannot be consoled. A broken, white-haired woman wobbles up the stairs and takes to her bed.

The show cuts to the present day and the story of Julia's ghost.

There are more reenactments: a security guard, once a "hardened skeptic," knocks on Julia's door and hears a woman's accented voice on the other side. "I'm in here," a German Jewish voice says, though no one is in the room. Another security guard sees Julia's face in the bathroom mirror. "I felt a cold chill come through me," he says, "and something told me it's time to leave." A hotel guest feels something staring at him as he lies in bed. He looks up and sees a white apparition—Julia, in her nightgown. The guest receives a psychological evaluation and a physical exam from the ghost hunter, to make sure he doesn't have a brain tumor. The hunter looks at the blueprints of the building; takes samples of the water, air, carpet, and wall paint; and tests for toxins that might have caused the illusion—phosphine gas can produce poltergeist-like flashes when mixed with air, he says. He brings in ghost-hunting equipment and seals off the upstairs of the mansion for seventy-two hours.

"We were able to analyze the environment in every possible way," the ghost hunter says, "and we did not encounter anything which was unusual or extraordinary." But he finds his witnesses credible. "Based on all the results of the investigation conducted here, we haven't found any facts to disprove the fact that La Posada is haunted," he says. "There certainly is that possibility that there is unusual paranormal phenomena taking place here."

This is how it is in ghost hunting: ghosts are present until proved absent. Absence of evidence, as they so often say in the world of the paranormal, is not evidence of absence. We so badly want the dead to stay with us.

And therein lies an industry. There are now a number of television "reality" shows that follow ghost hunters on their professional rounds. My favorite, *Ghost Hunters*, features two former

Roto-Rooter plumbers from Rhode Island. Their typical mission involves visiting a haunted property, speaking to the owners, and setting up the contemporary gadgetry of spiritual exploration— audio recorders, electromagnetic field (EMF) meters, Geiger counters, geophones, digital thermometers, and video, thermographic, and night-vision cameras—using machines to capture the ends of the visual and auditory spectrum where the dead tend to dwell.

The first episode I watched told the story of a Los Angeles waitress who had been found sliced in half in 1947. The team of five or six men in jeans and hooded sweatshirts, along with one woman, set up camp in a sprawling midcentury home where they suspected the waitress was killed. Some of the members sat outside at their laptops, monitoring the electronic activity; others wandered the rooms. Between night-vision flash cuts of blood, faces, orbs, and eddying smoke, the team members commented on the action. "We have a very interesting anomaly," they would say, or, "What the flip is that?" The monitors noticed a figure that looked almost like a human sliced in half. After many replays on SpecterCam Three, the ghost hunters determined that the blurry half ghost was instead a member of their team. There remained, however, some inexplicable voice recordings. Probably, they concluded, a haunting. So much equipment, so little closure.

Next, I interviewed a local ghost hunter in a sandwich shop near my home in Boulder, Colorado. Steve looked to be about forty years old. He had green eyes and a judicious brown goatee with a nickel-size patch of gray. His group worked almost exclusively on residential cases. They would start by sending two team members to interview the homeowner who suspected a ghost and to examine the floor plan, to "see what we're dealing with." This also gave them a chance to judge their clients——"we do get people with mental issues," Steve said.

Much of their work, he told me, involved hours and hours of sitting in the dark, waiting for something to happen, and then hours and hours of reviewing tapes to see if something had happened that was undetectable to the naked human ear or eye, such as an EVP (electronic voice phenomenon) or a strange video image. There were plenty of nights when nothing at all occurred. Sometimes they would find a rat in the basement or an improperly grounded electrical box. But there were also nights when they did uncover something they couldn't explain: a baffling moan, a drop in temperature, a spike on the EMF monitors. Seldom anything scary. "Actual demonic activity is rare," Steve told me.

This statement was reassuring to me, because, to be honest, I was not in the mood for demonic activity. I was, frankly, a little scared. The just-the-facts journalist in me thought of the whole ghost-hunting endeavor as sort of a joke——a punch line to my more meaningful historical search. But there was also a side of me that, faced with the prospect of spending the night in Julia's room, truly wanted Steve's counsel.

In truth, the world of ghost hunters and psychics was an unexplored frontier as strange and scary to me as New Mexico must once have been to Julia. I was terrified of the dark room and the long hours stretching in front of me alone in the night. I didn't know if I wanted to see Julia or not. My imagination regarding what I might find in Julia's room was vague yet vivid——I had seen the horror movies. I knew the undead could do ghastly things in the dark of the night.

Still, I doubted Julia would harm me——I was her blood, after all. And I had spent the past months combing through archives and family trees and the rambling corridors of the Internet, trying to understand who she was and where she came from. She would certainly have to understand that I came in peace.

According to Steve, though, I probably didn't need to worry. It was, he said, highly unlikely that Julia's ghost would appear on demand and provide answers to all the questions obscured by time and death. Or even that she would appear at all. When I told Steve about my plan to visit Julia's room, he advised me gently that I shouldn't get my hopes up. Steve had seen only one materialized ghost (head, arms, shoulders, no body) in his many years of looking. It was unusual to see a spirit on one's first ghost-hunting expedition.

On the other hand, my odds might be better. As a relative, Steve said, there was a chance that I could serve as a "trigger object"; I might induce Julia to appear in one form or another—light, sound, head, body. If she did turn up, Steve advised me to remain calm: there was generally only one big event per night, he said, and I might miss it if I lost my composure. I should be as steely as Abraham was with his revolver in the carriage with Sister Blandina, facing down a legend.

eight

❧ BRICKS AND MORTAR ❧

Archbishop Jean-Baptiste Lamy.

Sister Blandina had traveled to Trinidad with Julia and Abraham—and risked the ambushes of outlaws—because she was asked, or rather, obliged to do so by "the most Rev. Archbishop Lamy." This was Jean-Baptiste Lamy, the highest-ranking Catholic official in the Southwest—beloved, laureled, and feted throughout New Mexico. Why had he insisted that Blandina travel to Trinidad with the Staabs?

There was a connection between my Jewish family and this famous Catholic archbishop—a rather strong one, it appeared. How strong, exactly, was a question that had nagged at me for many years.

Lamy is famous in New Mexico, even today, because of the ambitious cathedral he built at Santa Fe's heart, a few hundred feet from the home that Abraham would build for Julia. But Lamy is known more widely in the world of literature because of the Willa Cather novel based on his life, *Death Comes for the Archbishop*. In the book, a Lamy-like priest named Jean Marie LaTour arrives in New Mexico in 1851—the same year the real Lamy arrived, and only five years before Abraham did. To get to New Mexico, LaTour endures a shipwreck off the Texas coast and an almost fatal wagon rollover near San Antonio—just as the real Lamy did. LaTour then survives a near year-long odyssey through the southern desert between Santa Fe and Durango, a horrific, fractured land of deep canyons, "the very floor of the world cracked open"—to request a letter from a bishop confirming his assignment. The real Lamy did this as well.

Lamy was French, born in 1814 in a clay-plastered house in the south-central plains of Auvergne, the son of well-to-do burghers. He was tall and lean, and very handsome, with a strong, square jaw and dark waves of hair swept back from a broad, contemplative forehead. He was gentle—nicknamed "the Lamb" as a schoolchild—and inclined to fits of bad health "not always entirely physical in origin," according to his biographer Paul Horgan. There was, said Horgan, a "nervous fragility" to Lamy, though in her novel Cather describes a quiet power in the man. "A priest in a thousand, one knew at a glance," she wrote of the hero whom she based on Lamy: "brave, sensitive, courteous."

His manners, even when he was alone in the desert, were distinguished. He had a kind of courtesy toward himself, toward his beasts, toward the juniper tree before which he knelt, and the God whom he was addressing.

It would take such a man to tame the vast, inaccessible parish Lamy had inherited. It was in utter disarray when he arrived—fractious, venal, muddled, and disregarded. Many of the churches were in ruins, built of dust and returning to dust. There were only nine active priests in a diocese that covered two hundred thousand square miles, and those priests took their vows lightly. They drank to outrageous excess. They gambled, danced, wore dirty vestments, and threw fandangos. They ran general stores, lived in open concubinage, and reared entire families of illegitimate children. They charged—and pocketed—exorbitant fees for baptisms, marriages, and burials, even for simply preaching once a year to a far-flung congregation. The result was that thousands of men and women who considered themselves Catholic lived unbaptized, unconfessed, unconfirmed, unmarried (though living in sin), and unforgiven.

Catholicism in New Mexico had a different flavor than it did in France. There was a theatrical element in New Mexico—the gaudily decorated altars, waxen priest dolls, and weeping, ring-kissing congregants. The statuary was vivid: bloody, "agonized Christs," as Cather put it, "and dolorous Virgins." The priests there had been left to their own devices for two and a half centuries. It was a daunting task to bring European piety to such an impervious parish. But Lamy was an ambitious priest—"Providence seems to have fitted me for a barbarous and extensive mission," he wrote—and he rode horse- and muleback the length and breadth of his diocese, building churches, suppressing heretics, restoring the celibate priesthood. He imported French priests as his deputies, as well as nuns to tend to the sick and the poor.

In time, Lamy came to believe that New Mexico also needed a more permanent symbol of the Roman church's renewed authority:

a proper cathedral. There were humble adobe chapels throughout the territory; they were endearing, thick-walled constructions. But Lamy lamented their "poor fabric of mud"—straw and dust, so primitive and impermanent. They reminded him of poverty, of barnyards. The churches of Lamy's youth had been substantial, sober constructions of dark volcanic stone, rounded and shadowed, with thick columns and heavy arches. He envisioned the same for the church he planned to build—a house of God in the Romanesque style. This new cathedral would be defiantly *not* adobe.

Lamy scouted materials. He located a cache of ocherous limestone in the Arroyo Sais that ran through town. He came upon a light volcanic tufa, found in the vaults of the Cerro Mojino just outside Santa Fe. These materials, laboriously quarried and hauled and cut and laid stone by stone, would form the cathedral's walls. A French architect and stonemason, Antoine Mouly, would direct the work.

The cathedral's granite cornerstone was laid in 1869, three years after Julia's arrival. The stone had been carved out to enclose a time capsule that contained the names of the president (Ulysses S. Grant) and the territory's governor, along with newspapers, documents, and coins of gold, silver, and copper. But the capsule disappeared a week later—stolen for the coins.

Such was Lamy's luck throughout the construction. The foundations were laid incorrectly, and they had to be torn up and started again. Funds ran abortively short, and Lamy's grand architectural ambitions had to be scaled back. Mosaics and carved figures were jettisoned. The nave grew smaller, the tower shorter. The structure, built on an old grave site, kept settling until Mouly was forced to add a new and costly set of subordinate arches to its sides. Then Mouly went stone-blind from the dust. Construction ceased entirely between 1873 and 1878, even as Lamy was elevated, in 1875, to archbishop. He raffled off his horses and his carriage in hopes of bringing in more funds. He begged the wealthy families in the parish—the *familias* Sena, Contreras, and Perea—for more money.

And finally, when their generosity was expended, he looked elsewhere—to the Protestants, and then to the Jews.

◇◇

As the archbishop struggled to complete his life's work, Abraham contemplated his own building project.

By the late 1870s, Julia and the children had returned from Germany. The children were at easier ages now, their characters defined. Bertha, my great-grandmother, was flirtatious; her older sister Delia was forceful, Arthur willful, Julius sweet; Teddy, the youngest, was impish and playful. Julia began to circulate in society more, attending parties and other gatherings.

She had probably learned English by then, though the family still spoke German at home. Perhaps she learned Spanish as well, though she would have less need of it soon. The railroad was approaching, and more Anglo men and women were arriving each day. Some floated through, but others stayed. Santa Fe was now on the verge of joining the larger world. Brick by brick, railroad tie by railroad tie, Indian battle by Indian battle, it was being transformed from a foreign territory into an American outpost.

In 1880, after many delays and much drama, the railroad finally arrived. Two years earlier, the Atchison, Topeka and Santa Fe Railway had announced that the main transcontinental line would bypass Santa Fe—notwithstanding the "Santa Fe" in the company's name—and pass through Albuquerque instead. The local business community was devastated, and a group led by Abraham and Archbishop Lamy campaigned to pass a bond issue to aid in the building of a spur from the main line to the city. On February 16, 1880, the first train puffed into Santa Fe. The Ninth Cavalry Band, part of a Buffalo Soldier regiment,

led a flag-waving parade of soldiers, carriages, and students from the
Plaza to the depot, where the territory's governor, its chief justice, and
Abraham drove in the silver spikes. President Rutherford B. Hayes
paid the city a visit later that year, the first presidential visit to the territory. Abraham served on the welcoming committee, and his brother
Zadoc, visiting from New York, rode in the president's coach from the
train station.

The city was growing more civilized. In 1881, the first streetlight
winked on, and a "new gasometer and conduit" was erected to light the
Plaza and the nearby streets. The first telephone line arrived in New
Mexico around the same time. There were now "fresh oysters daily"
at Miller's, according to ads in the *New Mexican*—mollusks, hauled far
from the ocean. Abraham built a plank sidewalk in front of his stores
so customers didn't have to slog through dust and mud; other establishments did the same.

Now that Santa Fe was an American city, it was time for Abraham
to build a mansion befitting his American dreams. He would build it for
himself, certainly—but also for Julia. It would be a proper European
house; one that might, finally, make her feel at home. He had vowed,
under the chuppah in Germany nearly twenty years before, that he
would provide for her. Perhaps he couldn't make her happy—perhaps
there weren't enough nuns or sisters available for that. But he could
build her a home. In 1881 Abraham purchased six acres directly east of
Lamy's growing cathedral on Palace Avenue—a fittingly royal street
name for this merchant prince's palace. Then Abraham imported, first
by steamer and wagon train and later by railroad, masses of pressed
brick, marble, and mahogany. His associates at the Atchison, Topeka
and Santa Fe lent him masons from Kansas City.

The house rose. Abraham's construction went far more smoothly
than the archbishop's. The *New Mexican* made regular reports on the
home's progress. "Mr. A. Staab is doing a good deal of building. Good

for him!" it reported. "The mansard roof of the new residence of Mr. A. Staab is nearly completed," came an update a few months later. In early 1882, a reporter took a tour. "S. B. Wheeler, the architect, showed the reporter through the splendid new residence of Mr. A. Staab. It is truly an elegant structure, doing credit to Santa Fe. . . . It is rapidly approaching completion." The house featured brass chandeliers, inlaid wood flooring, fluted door and window frames, and steam heat. The lower floors were conceived as receiving areas—two parlors, one for family, one for guests; a library; and a conservatory.

A mahogany staircase led to the second floor, which housed the bedrooms. Separate bedrooms for husband and wife were a luxury that Abraham and Julia could now afford. Abraham's lay to the left of the grand staircase. Julia's—the one I would visit a hundred and thirty years later—lay to the right, with its arched windows overlooking Palace Avenue. She would have her own bathroom, with a claw-foot porcelain tub. The children would sleep in the back rooms. In the third-floor ballroom, the Staabs would host elegant affairs—fetes and formal dances, no fandangos there. For contemplation of the vistas, Abraham topped the home with an elaborately decorated widow's walk.

It was a structure that taunted the land around it. Mastery! Triumph! Not for Abraham the becoming modesty of the native architecture. Not for him the blending of house and land. "The Hopi villages that were set upon rock mesas, were made to look like the rock on which they sat, were imperceptible at a distance," Willa Cather wrote fifty years later in *Death Comes for the Archbishop*. "None of the pueblos would at that time admit glass windows into their dwellings. The reflection of the sun on the glazing was to them ugly and unnatural—even dangerous. . . . It was as if the great country were asleep, and they wished to carry on their lives without awakening it; or as if the spirits of earth and air and water were things not to antagonize and arouse." But Abraham wasn't the sort to fear sleeping spirits.

Furniture came from Europe and the East Coast. A piano traveled from Kansas City. A mezuzah was affixed to the doorframe. Workmen installed gilt floor-length mirrors. On the grounds around the structure, gardens and a large orchard were planted. "A green sword now graces the residence of A. Staab," said the newspaper. My grandfather, who was a child when the family sold the house, remembered that the apricot trees reached over a tall wrought-iron fence, and that the local children would shinny up to pick them.

◇◇

Archbishop Lamy helped Julia plant those apricot trees, which long outlived them both. In the gardens at La Posada, one tree still stands, the one my children climbed—three thick limbs, gnarled and dendritic, braiding over and through the adobe wall and roof of one of the casitas. A plaque leans into a particularly large and impressive knot near its base: "In the 1880's," it reads, "this apricot tree was planted by Julia Staab and her dear friend, Archbishop Lamy. They were avid gardeners and together planted all of the other large fruit trees on the grounds of La Posada de Santa Fe."

Abraham and the archbishop were friends. They shared an interest in the civic improvement of Santa Fe, and they worked together to bring the railroad as well as the sidewalks and gaslights. They constructed monuments to their beliefs—a cathedral for one, a private mansion for the other—and strove, side by side, to impose a European sense of order on their adopted city. But it seems that there was a different kind of friendship between the archbishop and our displaced Jewish bride. It was a native sympathy, built on quieter tasks and more delicate sensibilities.

This was the relationship that Lamy's biographer Paul Horgan described in another book he wrote about the Southwest, *The Centuries of*

Santa Fe, published in 1956. In one chapter of that book, he tells of the friendship between the archbishop and an unnamed German Jewish woman. The chapter is titled "The German Bride," and its first page is illustrated with a drawing of a three-story Victorian home, surrounded by deciduous trees and a tall wrought-iron fence. It is the precise image of Abraham and Julia's home.

The German bride was, in Horgan's depiction, an exquisite and dignified creature in a rugged outpost starved for urbanity. My family loved that bride, who seemed to have floated right out of a Western. "Her skin was white," Horgan wrote. "Her clothes were beautifully made in the highest of fashion. She animated them with something of the effect of a small girl dressed up playing queen. She could make everybody smile simply on meeting them. Wait till she played the piano for them, and then she would make them sigh, or even weep. Her Mendelssohn—they would never believe it."

The German bride was a consummate hostess, as Julia might have been on her good days. Horgan describes elaborate formal affairs in the bride's mansard-roofed home, and afternoon teas in the mansion's yellow-silk drawing room, and dinners at a table set with "European china, cut glass, silver, lace, and linen." There were visits from Rutherford Hayes; Generals Nelson Miles, Philip Sheridan, and William Tecumseh Sherman; and once, the "notorious philosopher Robert G. Ingersoll"—a famous agnostic.

The bride entertained the archbishop on a regular basis. "She would find in him a friend," Horgan wrote. The two seemed to understand each other, their "distinct sophistication" and European sensibilities. "She always enjoyed her little exchanges with the bishop . . . ," Horgan explained.

Her education was excellent, and she spoke a social kind of French, so that when they met she engaged the bishop in his own early lan-

guage. He replied in kind, amused to speak the language in which he
still realized much of his thought.

Abraham had left Germany of his own accord, willingly, forcefully.
Julia and the archbishop had arrived in Santa Fe under higher orders—
Lamy's from the church, Julia's because of the husband she had sworn
to obey. And the longing for home never left the archbishop or the
German bride. They never quite adapted, in Horgan's estimation, to
the high seasoning of the food, or the high drama of the landscape and
people around them. Lamy and Julia were both avid gardeners. Lamy,
accustomed to the innumerable greens of the Limagne Plain where he
had grown up—the yellow-green grasses, the silvered willows, the
near-black hearts of the poplar stands—never grew to love the desert
reds and buffs and taupes and tans. Julia, too, favored the gentler, more
generous blooms of her childhood home.

When Lamy first came to New Mexico, he carried cuttings from
France, and each time he traveled to Europe, he brought back more—
peaches, pears, oxheart cherry, fall and winter apple; seeds of cab-
bages, turnips, and beets; muscat and Malaga and Gamay and Catawba
grapes hauled in buckets of water across the ocean and the plains.
He planted them behind the parish church that would be replaced,
eventually, by his new cathedral. The garden was his only personal
indulgence—his only visible one, anyway. It was five acres, an adobe-
walled garden with a fountain, a sundial, aisles of trees, formal walks,
shaded benches, and a spring-fed pond with water lilies and trout. He
brought shrubs and vines and shade trees with him, too, thousands
of them, chestnuts and elms, locusts and osier willows, that he transplanted along the Plaza and the streets that radiated out to mountain
and desert.

Some of those cuttings also found their way into Julia's garden.
Transplanted themselves to an odd and barren land, Lamy and Julia

performed their own acts of reclamation, irrigating those things that couldn't survive without intervention, softening their new city's stark splendor. Daguerreotypes from the 1850s show a dusty stretch of plaza, bereft of vegetation. But by the 1880s, there was bountiful shade from the trees Lamy had imported. This desert did not grow green on its own; it required nourishment. In Lamy's hands, even the most delicate varieties flourished.

Nor was gardening the only affinity between the archbishop and the German bride. There was the love of European architecture, the conversational French. They were also both often unwell. Julia's mental and physical health was tenuous, as we know; the archbishop, too, was "always ill," according to Horgan. He was bled twice, Horgan reported, and treated fifteen times with leeches on the abdomen. His mental state also seemed incongruously fragile "within his square peasant frame," Horgan wrote. Lamy was nervous; there was a darkness within him. He had a tendency to collapse into himself and withdraw from the world. Of course, Julia did as well.

There was a kinship between the archbishop and Julia—a connection.

nine

❧ OTHER SPECULATIONS ❧

Grandma Ginny.

My grandmother Ginny was once a new bride in New Mexico, too, and she also speculated on the relationship between Julia and Archbishop Lamy. Ginny was my sole non-Jewish grandparent—a Westchester County WASP who met my grandfather at a Yale football game in the days of raccoon coats. After they married in 1935, he drove her to her new home in New Mexico. In an essay she wrote called "The New Bride," she compared Julia's experience as an imported wife with her own. In the essay, she describes her arrival in Albuquerque for the first time, and feeling violently out of place in that desiccated land. The road signs on Route 66 had advertised "the promised land for 500 barren miles," Ginny wrote, "and sign by sign I envisioned a beautiful oasis on the banks of the fabled Rio Grande."

But then the pavement turned to gravel, and Ginny saw Albuquerque in all its taupe and stony severity. "I was appalled," she wrote, "to put it politely." She had never seen such dust. Ginny was an up-to-the-minute woman, in knee skirts and shoulder pads, finger-waved curls crimped below her ears; such fashions hadn't yet arrived in New Mexico. At a tea party given in her honor, she was told "in no uncertain terms" that she should wear a long dress. "I rebelled but finally accepted the inevitable. I scraped to the bottom of my trousseau trying to find something suitable, but nobody at B. Altman's had foreseen that I would be expected to wear a long dress . . . when the temperature was 101 degrees."

All went well, Ginny wrote, until the plates were cleared away and the men lit up their cigars. "There wasn't a breath of fresh air, the red velvet drapes were closely drawn, the temperature still hovering around one hundred, and the spoiled New Bride, never having been exposed to such heat or to men who smoked cigars," plunged out the front door, she wrote, "and vomited quite thoroughly, trying not to spatter my dress or shoes. Between heaves I hung on to the trunk of a small sapling." Like Julia, Ginny was a woman of sensitive constitution stranded among men of business and left to her own sometimes insufficient resources. It was not easy to be a new bride in New Mexico—not in 1866, not in 1935.

Nor was it easy to be a mother and aging wife so far from home. I like to think that there was a time when my grandparents were happy—but all I know for certain is that in the end they were miserable. After thirty-five years, the marriage dissolved in a noxious stew of alcohol and anger. My grandfather remarried the day after the divorce came through, and Ginny retreated, at age sixty, to the brackish, bug-stippled pond in Rhode Island where she had spent summers as a child, and where she lived in a house on stilts. She vacationed on cruise ships, drove a Buick convertible, and maintained an excellent suntan. She drank coffee with her pinkie pointing skyward; at noon she switched to vodka. Ginny felt that she had lost much to those long decades in the desert—her youth, equanimity, and good humor. She never remarried. After her death we found a raft of poems that suggested there had been a great love in her life ("I write to you so often / In letters never shown") and that he was not my grandfather.

No wonder Ginny's writing about Julia seeped disappointment and longing. "I often wonder about the German Bride," she wrote, reflecting on Horgan's chapter on Julia. "Was she always so gracious and charming, the perfect lady? Wouldn't she have been annoyed by the rasping of frontier fiddles, condescending toward the provincial theatricals, bored with the literary pretensions of the exclusive Ladies Reading Society?" Ginny aired "other speculations" about Julia, as well. "Did she really continue to adore with undeviating devotion her entrepreneur husband who got her into this mess in the first place?" she wrote. My grandmother clearly hadn't continued to adore her own entrepreneur husband.

Ginny had written this essay before the family learned about Julia's ghost, but she, too, projected onto Julia her own preoccupations. For me as a younger woman, Julia's story was about women professionally and politically oppressed; for Lynne and Joanna, it involved cruel husbands. For Ginny, I suspect, it reflected the horror of marrying the wrong person—and then learning, perhaps, that the right one was nearby, yet painfully out of reach. "When a young French priest arrived," Ginny wrote about Julia, "the young bride and the young bishop had much in common."

The properties of The Mansion, her residence, and The Manse, his, adjoined. They had walks and talks, always in cultured French that nobody else could understand, between the two establishments. She would admire his gardens and flowers, he would inspect his orchard, the pears and peaches and apricots that he had imported from his French homeland. He would pick the perfect ones for her to sample and they would rest under an arbor of his favorite grapes while he recited classic French poetry.

My father and I used to joke, after Ginny died and we read through her trove of papers, that someone should write a romance novel about a German bride and a smoldering, chestnut-haired French priest. It would be a tale of ripped bodices and vestments, expensive perfume on clerical collars, and it would be called "Love Comes for the Archbishop."

It was enticing to think of Julia as a woman of hidden passions. Perhaps, we speculated blithely, it wasn't the plucky, bulldoggish merchant who sired our line, but instead the sinewy, blue-eyed, square-jawed archbishop of delicate sensibilities and posthumous literary fame. Maybe, we joked, we were a little bit French and a little bit Catholic, products of a peculiarly Western ghost story that was also a great love story.

There are, of course, many—many—reasons *not* to believe that the archbishop was our forebear. To start with, he was a priest. In New Mexico at the time, that did not rule out fathering an octet of children. But Lamy was not, by anyone's accounting, dissolute: he fought hard in his early years as bishop to enforce the celibate priesthood among his fallen brothers in New Mexico. He was driven, moral, "a priest in a thousand." Then, too, he was quite a bit older than Julia—thirty years. So he wasn't exactly the "young French priest" that Grandma Ginny had imagined. His days of grand passion (if he'd ever had such) were likely behind him when Julia arrived, and the long years in the desert were also not easy on his Fatherly good looks—"That country will drink up his youth and strength as it does the rain," Cather wrote. And it did. By the end of Lamy's life, he had lost his teeth, and his face had sunk in on itself—not the stuff of romance novels. But still!

Partly on the strength of Grandma Ginny's speculations, and partly to satisfy my own curiosity, I decided to get a DNA test. Because unlike ghost stories, some love stories can be confirmed in the world of hard science. It is as easy as swabbing a cheek to learn of four letters, ACTG—the four base molecules of DNA—configured and reconfigured through blood and bone, past and future, to tell us who we are and where we came from.

I did some research, got online, added one (1) DNA kit to my cart, pressed Submit, waited for the kit to arrive, waited two hours after a meal, swabbed my cheek, swabbed it a second time, placed the swabs and my chromosomes in two plastic vials of liquid preservatives, wrapped them, packed them, and mailed them off to Texas.

Lucy

◇◇

WHILE I WAS WAITING for my DNA results and working up the courage to spend the night in my own haunted hotel, I decided to sharpen my ghost-hunting savvy by visiting another famous spot. I drove from my home at the base of the Rockies up a narrow river valley that opened into a mountain-ringed basin, and pulled into a parking lot warm with afternoon heat. Before me, at the foot of a gigantic lumped-granite hillside, loomed a red-roofed, white-sided neo-Georgian colossus—the Stanley Hotel of Estes Park, Colorado. The hotel provided inspiration for the author Stephen King's famous horror novel, *The Shining*, and it is said to have a head-spinning diversity of ghosts. F. O. Stanley, the hotel's long-dead original owner, paces the lobby. His wife, Flora—the Julia of the establishment—plays piano in the ballroom. There is also a man in a dark gray suit, and a charred chambermaid, and a nanny who tucks in unsuspecting guests. Invisible children run down the hallways. Full-blown ghost parties rock the ballroom.

The Stanley runs nighttime ghost hunts each weekend in summertime; I had signed up for a Friday-night tour guided by two hotel-employed paranormal experts, Connor and Karl. Connor was blade-thin with downy blond hair, Karl more robust and jocular. We began with a briefing at a gazebo near the hotel's front porch. There were thirty of us: a fair number of buzz-cut teenage boys, a group of ample middle-aged Nebraskans, and a handful of girls with headbands and flip-flops.

Connor and Karl began by laying out some rules: Keep our cell phones in airplane mode. Minimize talking and whispers.

If our stomachs grumbled, confess. ("It can sound like people screaming, women talking, demonic growls," Connor said, "really!") Karl and Connor couldn't promise us a ghost, but they assured us they had seen plenty. "We've heard voices from nowhere," Karl said, "heard footsteps, felt pants tugged, seen cell phones fly a few feet across the room. Sometimes a ghost will sit in someone's lap."

Connor told us that the Stanley's guests most frequently encountered "residual" hauntings—historical memories that ran over and over, like a filmstrip: tinkling pianos, children laughing, the groans of trunks being dragged across the floor, smells of tobacco and rose oil, lights, shadows, cowboy silhouettes. Residual ghosts, Connor explained, have no idea that they're dead. "Intelligent ghosts," on the other hand—ghosts like Julia—can interact with people.

Connor then mentioned a type of ghost encounter called a "timeslip," in which the present grazes against the past. Imagine two clotheslines in a row, he said. "Most of the time, the sheets hang down and are still, but maybe there's a metaphorical breeze, a brushing between the sheets." When the sheets touch, he said, people who don't live in the same era see each other as ghosts—strangers in period dress. You peer briefly into their life; they peer into yours.

I liked this idea. It was exactly this that I was trying to do: riffle through the residua of Julia's time on earth in the hope that I could generate my own breeze and look into her life. What I had seen, so far, was a sad bride who longed for home, who had no choice in what her life would be. I wondered what she would see if she peered into mine.

Next, Karl and Connor engaged in some polite debunking. Not every weird sensation portends a ghost, they explained.

Carbon monoxide can cause visual hallucinations. So can mini—epileptic seizures. Sounds that fall just below the human range of hearing can cause humans' eyeball liquid to vibrate, making us see fleeting images out of the corners of our eyes. Our brains can play tricks, too, especially in that borderland between wakefulness and sleep, when reality blurs into dream—called a "hypnagogic state" when falling asleep, a "hypnopompic state" when waking up. We can also will ourselves into believing ordinary creaks and gusts are phantoms. That's called apophenia—the tendency to see meaningful patterns in meaningless data.

Connor and Karl pulled out two metal suitcases full of gadgets that were supposed to help us avoid these kinds of mistakes: flashlights, a radiograph, an infrasound detector, a seismograph, a static detector, and an EMF "K-II meter" intended to pick up electrical frequencies. As dusk fell, they split us into two groups; mine would start in the main hotel building with Connor. We climbed four flights of broad, winding stairs and filed into a small room with a king bed, a sitting porch, an antique clock, and photos on the wall of women in turn-of-the-century, all-white dress. This was room 401. It was probably the most haunted spot in the whole hotel, Connor said.

We crammed in, thirteen of us, perching on the bed and leaning against the walls, and Connor turned out the lights. Blazes of lightning from a brewing storm burst intermittently through the gauze curtains, flickering on Connor's face. He placed a seismograph on a shelf in the closet, then turned on a static detector, which looked and behaved like a spinning top; he then handed out a batch of flashlights and EMF detectors.

One of the teenagers, overweight and with fuzzy, dark hair and glasses, took his detector and smartphone into the bathroom. "EVP Session One," he said in a slightly froglike voice.

He was hoping to capture an electronic voice phenomenon or an electromagnetic field—anything, really. "I got a spike in the bathroom!" he yelled. Others followed him into the small space; a traffic jam ensued. Connor told the rest of us that he suspected the room's ghost was the evil Lord Dunraven, a brothel owner and land-grabber who had been run out of town in 1907. Lord Dunraven probably resided in the closet. "Lord Dunraven," Connor asked, "what color is my shirt?"

They didn't allow ghost hunters or tours into Julia's hotel, out of respect; now I could understand why.

After a few suspicious radio squawks and flashlight blinks—nothing definitive, Connor told us—we transferred to the concert hall, where Karl would now be our guide. We wandered first through the dark entry hall and up to Flora Stanley's private viewing box. "You often get a scent of roses here," Karl told us, though I smelled only aftershave. From there we descended carefully to a cellar room full of pianos—unnerving in themselves, those dark coffins of sound. "We'd love to chat for a little bit," Karl said to a ghost named Paul. Someone's stomach growled. It was mine, and it did sound rather demonic. We heard a thump—the wind, I suspected, as the storm whipped up the valley—then nothing. Karl told us of the time he felt a ghost pass through his skull.

He also mentioned that a spirit named Lucy often came to visit the basement bathroom. According to local lore, she had been a squatter in the early 1970s—a hippie—who was found and evicted and later froze to death. A group of us filed into the bathroom and sat down on the floor. We unscrewed, slightly, the top of a flashlight and placed it beside us—this was an old ghost-hunting trick. The flashlight sat dark for a moment.

Then it turned on. Everyone got very excited. The light went off, and on, and off again. "Lucy, if you're here, turn on the light," the EVP boy said. It didn't turn on at first. And then it did.

"Lucy, if you want us to stay, turn off the light," the boy said. The light remained on. We stayed anyway.

"What kind of music do you like?" asked the boy. Our guides had told us that Lucy liked music. "Pop?" Nothing. "Rock?" Nothing. "Country? Metal?" The light turned off.

"I think that was a delayed response to rock," said the boy. A matronly woman from Nebraska disagreed: "No, I think it was country." She turned to the flashlight. "I like country, too," she said, in what I suspect was her sweetest voice. The EVP boy couldn't believe that Lucy liked country. He wiped a bead of sweat from his forehead. "If you like rock, turn it back on," he said.

It went back on, then went off and stopped cooperating. Everyone tried to coax Lucy back, but the light stayed off. The EVP kid called Lucy a spoiled brat; the woman from Nebraska swooped in to defend her.

I collected my flashlight and K-II meter and left the room. I couldn't see what I was going to learn from this exercise. Perhaps there were ghosts here; perhaps Lucy really did lurk in the bathroom, debating her musical tastes. But this was no way to talk to the dead. This was no way to explore the bounds of our mortality. This was just stupid. I wasn't sure that I would be any more successful than these hotel ghost hunters in my parsing of the scant documentary clues that Julia had left behind. But it seemed a far better course than sitting on a bathroom floor talking to a flashlight made in China, hanging on every gust and grumble and flicker.

If I was going to find Julia, I would have to do it some other way.

ten

❧ FOUR LETTERS ❧

Santa Fe's Saint Francis Cathedral.

The dead hold secrets that we can never know—and in that respect, ghost hunting and ancestor hunting are not so far apart. They both involve sifting through heaps of supposition, extrapolation, and unmoored clues.

As I pored through old books and articles about the Staabs, I encountered another mystery regarding my family and the archbishop.

This was the matter of the cathedral itself, the construction of which resumed in 1878, once Lamy was able to raise additional funds, and continued steadily thenceforth. In 1884 a vaulted roof enclosed the new nave. In 1886, the bell tower rose above the city. Lamy had finally realized his dream. The cathedral is the brick-and-mortar ghost he left behind, visible everywhere one stands in downtown Santa Fe. The building is not as grand as the archbishop had hoped—the ceilings lower, the nave narrower—but it is still impressive, so much more substantial than the rumors and apparitions that constitute Julia's legacy. It is a sweet if stolid building, with Corinthian columns, a rose window, and rounded arches. But in the keystone of the entrance arch, Lamy commissioned something odd. Above the door, contained within a triangle, are four Hebrew letters, YHWH: the Hebrew word for God.

I had always been told that the archbishop commissioned those Jewish characters on his cathedral because of Abraham Staab. That's what my family told me growing up; that's what the history books said; that's what the tour guides in Santa Fe say. The story goes like this: While Lamy was struggling to complete the cathedral—that physical manifestation of his spiritual goals for an untamed territory—he attended a poker game with Abraham Staab and other local merchants, lawyers, and politicians. The archbishop never gambled, the storytellers insist, but he liked to attend the games for company and fellowship, and on this particular evening Abraham noticed that his friend seemed particularly subdued.

Abraham asked Lamy what was wrong, and the archbishop responded that he feared he would not live to see the cathedral completed. "Times are hard," Lamy said. He had run out of funds completely, he confessed to Abraham. Abraham didn't hesitate. He asked Lamy how much he needed, and the archbishop said it would take somewhere between ten and fifteen thousand dollars. That was a considerable sum in those days, but Abraham promised him the money immediately. He imposed only one condition. "Cautiously," wrote William Keleher in his history of New Mexico, *The Fabulous Frontier*, "the man of God measured the eyes of the man of Commerce and Business and inquired: 'To what extent, how, Mr. Staab?' Staab replied: 'Let me put one word above the entrance of the Cathedral, chiselled in stone.' 'And what is that word?,' parried the Archbishop. 'You must trust me, Archbishop,' replied Staab." And thus, YHWH, Yahweh, the letters Abraham requested, were chiseled over the entrance of the cathedral like a brand.

The letters are still there, carved into their stone triangle above the cathedral door. I had seen them many times as a child, but I hadn't visited in recent years. So on one of my trips to Santa Fe, I went again to look, swept first into the nave with a gaggle of tourists on a hot August day. The cathedral was smaller than I remembered and a bit dark, with a large painted altar that reached almost to the top of the vaulted nave, rows of saints—the risen dead—stacked up to the ceiling. Burgundy-patterned hotel-style carpeting ran up the center aisle. Outside, though, little had changed since the archbishop had supervised the cathedral's construction over a century before. The tawny sandstone glowed against a cloudless sky, and the Hebrew letters overlooked all who passed through. I listened as a tour guide told once again the story of Abraham and the archbishop to a visiting group.

This version was a slight variation on the one I had heard when I was younger. In this one, there is no mention of poker. Instead, the archbishop comes to Abraham's office asking for an extension on promissory notes for money that Abraham has lent him for the cathedral. Abraham pulls the notes from his safe, tears them up, and throws them in the fire, forgiving the debt. My great-great-uncle Teddy, however, stood by the poker version, which he recalled hearing many times from his father—though he averred that "under no circumstances" would his father have traded money for the privilege of having Hebrew letters on the cathedral. "He did not bargain with the highest religious officer of the diocese," Teddy wrote in a letter to a local historian. Abraham did it simply because he was a good man, Teddy said, and Lamy wanted to thank him, placing the letters on the arch unsolicited. Another version, told to me by a third cousin, holds that the archbishop frittered away the cathedral money because he had a gambling problem.

Then there is the version told by Floyd Fierman, the historian and rabbi from El Paso, Texas, who wrote extensively about New Mexico's Jews. In 1962, Fierman explored the letters between Lamy and his mentor in Cincinnati, and found no reference to "any communication with any known people of the Jewish faith, indicating a loan or a gift." Given the lack of documentary evidence, Fierman began to wonder whether the story of Abraham and the cathedral wasn't just "another of the legends that grows with such ease in the parched earth of New Mexico tradition once it is irrigated with the moisture of the lips and the tongue."

As Fierman dug further, he learned something else: Lamy's Hebrew letters were no Judeo-Southwestern novelty. They were enclosed in a triangle, and "in Europe," he explained, "this was a common Christian symbol." It was called a tetragrammaton, and it had been carved in numerous Gothic and Romanesque churches throughout northern Europe, including Lamy's native France. There was even a tetragammaton in the cathedral in Saint Louis, where the Santa Fe Trail commenced. "It . . . could be, once the emblem was carved," wrote Fierman, "that these Jewish friends, totally ignorant" that the Hebrew letters on the cathedral were not unique, "were actually pleased and did consider it a friendly gesture by Lamy! Which is all to the good in this world of strife and misunderstanding among peoples." The story of Abraham and the cathedral was, in his opinion, merely a legend—a ghost story.

I hadn't known, when I'd embarked on this hunt for my family's ghosts, whether Abraham was a good person or a bad one—or some combination of the two, as most of us are. But I'd known, because of the cathedral story, that he had at least been a generous friend; he had, for the small price of four Hebrew letters, helped to pay for the cathedral. This was an established fact. Except that now, it wasn't.

Fierman's article confounded me. The story of the archbishop had been handed down through my family as gospel; no one had imagined it was anything but real. But it was no truer, Rabbi Fierman suggested, than the ghost stories my family had always dismissed.

Still, everybody else believed the story of Abraham's gift to be true. In 1967, Teddy, Abraham's only living child (he was ninety-two at the time), received an award on his father's behalf from the National Fellowship of Christians and Jews. The archbishop of Santa Fe, James Peter Davis, gave Uncle Teddy a scroll in tribute. If the story of Abraham's gift was inaccurate, everyone was perfectly happy for it to be that way.

What tales can we believe? The submersive force of history—the sedimentary layers of narrative—seems to bury even the hardest facts, and only the physical clues jut above the surface: the hotel; the cathedral; the chiseled Hebrew letters; the apricot trees; and also DNA, those four letters that can peel open, tetragrammatically, our genetic past. These artifacts—a tantalizing few—are all we can trust. We see them, collect them, and try to grasp what they mean.

◇◇

It's not as if I expected my DNA results to support the hypothesis of my descent from the archbishop. No one besides Grandma Ginny—and my father, and myself, and a few conspiratorial cousins—had ever suggested there might be more to Julia's friendship with Lamy than what was decorous. I loved the drama of the suggestion—hidden passions, forbidden love. I delighted in the idea that my heritage might be mixed up in Santa Fe's own mélange of cultures and history, that I belonged to more than one tribe, and might be able to claim a famous French archbishop as my forebear.

But I didn't believe that he was. And it was easy enough to disprove, anyway: If my DNA test confirmed that I was, in fact, three-quarters Jewish, then it would be clear that the archbishop couldn't have been my great-great-grandfather. If I were descended from the archbishop, I would be only 68.75 percent Jewish.

When the results came in, I went to a web page where I opened up a bar graph with orange and blue bars representing my "Middle East (Jewish)" and "European" heritage. As I'd expected, the orange Jewish bar was the larger of the two. But it was not as large as it should have been. A second test confirmed: I was 68.75 percent Jewish—missing 6.25 percent, which is exactly the proportion of one's genes that each great-great-grandparent bequeaths. The results from each test also suggested that I was part French.

French. Nobody had ever mentioned French ancestry. My mother, who had been tested already, was 100 percent Jewish. Thus the great-great-grandparent-sized deficit of Jewishness came from my father's side—perhaps from a certain non-Jewish French archbishop.

Perhaps the French heritage came from other missing ancestors. There had always been conversions, affairs, rape, and intermarriage, even before the modern era. I could have tracked down more relatives from various branches of my father's family, insisted they get swabbed, and compared their ethnicities with mine—but short of an exhumation, questions would remain.

Stories and reminiscences had provided no hard certainties about the cathedral and Abraham's generosity. Nor could the tools of genetics shed any real light on Julia's relationship with the archbishop. And even the supposedly objective documents of history—books, letters, artifacts—were beginning to confuse me. Online, I found a few of Abraham's passport applications. In 1902, he swore, under oath, that he had arrived in the United States in May 1857. In 1906 he stated, also under oath, that he had arrived in 1856. Elsewhere he placed his arrival as 1854.

The more I dug, the less I knew.

◇◇

Still, Julia and Abraham did live once. They slept and woke, touched and tasted. They were there in the past; their traces could be found in the archives. They existed. Every time I saw the Staab name in a newspaper, in a ship's log, or in the index of an old book, a chill scuttled up my neck: the dead came alive for a moment.

It was Lynne, the genealogist who dreamed of Julia's death in the bathtub, who helped me retrieve Julia's brothers from those records: Bernhard and Benjamin Schuster, crossing the ocean, arriving in Santa Fe, living in the house on Burro Alley and working for Abraham. She tracked them from there to El Paso, where they opened their own dry goods business. They were well respected there: civic leaders, businessmen.

But Lynne also found a sadder story from Ben's time in Santa Fe. He had, while living with Abraham and Julia in 1879, fathered a child with a local Hispanic woman, Damasia Chavez. Lynne found a great-great-great-granddaughter of Ben's illegitimate daughter, Josefita, who told us that Ben had wanted to take Josefita to Europe with him. In the end, however, he hadn't, and he had gone off instead to found his own business in El Paso.

Lynne was convinced that Julia was devastated by this. "I would imagine that having Ben around Julia was a comfort to her," Lynne wrote me. "One of the classic behaviors I imagine for Abraham was that he wanted to control everyone (especially who his children married). . . . He would NOT want his sons and daughters to think bastard children would be tolerated. Ben may have been banished from the Staab household . . . maybe."

Maybe.

I wasn't sure that Abraham was quite the ogre that Lynne believed him to be, but the story made some sense. It is true that Ben didn't marry Damasia, whatever the reasons. He moved to El Paso—in shame, perhaps?—the same year that the baby was born. Such things as an illegitimate child didn't ruin a man, however, the way they did a woman. He still came to Santa Fe regularly—"Ben Schuster, of El Paso, hale, hearty, jovial and energetic, is in the city," wrote the *New Mexican* after the baby's birth. In 1883, he married a German Jewish woman, Sophia Berliner.

That was the year after Julia moved into her new mansion; the year before the archbishop's nave was enclosed. The archbishop was in declining health, as he wrote in a letter to the bishop who would replace him. "Not only my memory, but also my other mental faculties have much declined," he wrote; "the smallest serious effort, worries, cares, difficulties, exhaust me and make me ill." Julia was pregnant then— perhaps with the ailing archbishop's child, but more likely with Abraham's.

In July of that year, eight years after her last successful pregnancy, she gave birth to a daughter. "Mr. Abraham Staab has a little heiress at his home," the *New Mexican* reported on July 25. Julia was thirty-nine years old, and this was her final baby—the one who died and, according to the ghost stories, turned Julia's hair white, the baby who might have been drowned in the bathtub and had so horrified the phone psychic Misha—the child of the darkness. They named her Henriette, after Julia's mother.

Her life was woefully short. On August 9, 1883, the paper relayed the sad news. "Mr. and Mrs. Abraham Staab's home was made glad only a few weeks ago by the coming of a new life and a new joy, a tiny girl babe. To-day He who gave it called it home to His 'mansion in the skies not built with hands, eternal in the heavens,' and sorrow reigns instead of joy. The infant daughter was but three weeks old. Its death occurred at 8 o'clock this morning and at 8 o'clock this afternoon its white robed form was laid away in the Masonic cemetery amid the tears and regrets of the sorrowing parents and their large circle of friends."

In the prayer book, Julia wrote a new entry: "*Henriette A. Staab geb am 22 Juli 1883, gestorben am 9th August 1883.*" The writing is smudged and uneven, slightly off-kilter, contrasting starkly with the thick black letters listing Henriette's brother Teddy just above. It is written in pencil, as if Henriette's spirit passed through too quickly to warrant a pen and pot of ink. In some traditions, it is considered a blessing when a child dies before it can commit any sins, for then it is guaranteed a spot in heaven.

And it was at this point, the ghost stories explain, that Julia came unraveled.

Sarina

◇◇

I DROVE NORTH FORTY-FIVE minutes from my home to visit a psychic named Sarina. She met me at the door of a squat one-story brick office building in the old square-built farming community where she lived. Sarina was in her forties and wore racing-striped sweatpants and a flowered T-shirt. Chaotic brown curls unfurled across her shoulders. After we introduced ourselves, she sat down in a low-slung gray tweed armchair against the wall. I chose a rocker across from her. I didn't give her much information up front, only that I was looking for my great-great-grandmother, who was rumored to be a ghost.

Sarina looked into the middle ground off to her right. "I'm seeing her as older, in her sixties," she said.

I explained that Julia had died at the age of fifty-two. "That's interesting," she told me. "She feels that she didn't have the opportunity to complete things."

The February sun angled through slatted shades onto Sarina's face. She had come recommended by a friend, who had seen her many times and had great faith in Sarina's ability. "I see her with this big hat," she told me—a hat of the sort you'd wear in a play, or maybe to a play. "Did she appreciate or participate in the arts? She's dressed up, really dressed up. OK—and flowers, there's lots of flowers around her, I feel that she really loved flowers," she said. "To your great-great-grandmother it feels like flowers were a big deal. And fragrances, I smell the fragrances of the flowers."

Sarina smiled at the air off to her right. "She's going back to her younger days," she said. "The suitors were definitely plentiful. There were a lot of men who pursued her even though she was married." Perhaps the archbishop? Sarina couldn't say. But she did say that at some point Julia's "social butterflying" stopped, and her world contracted. A man made it so—it may have been Abraham. "She was asked to become more of a home person at that point."

Not that Abraham held her back, Julia told Sarina—except Sarina was feeling that he did, in fact, hold her back. How odd, I thought, to watch Sarina take issue with the air. But Misha and Lynne had said the same: Abraham stymied Julia. "She doesn't want to blame her husband, she wants to make that very clear, but she does want you to know there was this other part of her."

Sarina stopped and listened for a time. Julia had thrust an image into her mind in the same way someone might move to a new screen in a web conference, symbols floating in from the ether. "She's saying, 'One child,'—OK, then tell me about the one child. She said she lost a child."

And here my skull and neck grew cold, because I hadn't told her that Julia had lost a child.

Sarina had also lost a child. Her son was named JT, she told me, and he had died at age seven of the flu—not because he was immune-suppressed or otherwise unwell, but because horrible things just happen, sometimes, for no apparent reason. JT developed a fever, and five days later he was dead. She hadn't known she had mediumistic skills until afterward, when JT came to her and said that this was her calling: to serve as a bridge between children who had died and the parents they left behind. JT became Sarina's spirit guide. I remembered then that the friend who had recommended Sarina had, like Julia, lost an infant very soon after childbirth. Sarina specialized in bereaved mothers.

Sarina started to speak, then stopped herself. "I'm trying to put words in her mouth and she's saying, 'Don't say that!'"

She paused. "Obviously, that"—losing baby Henriette— "was a huge event, but that's not when things shifted. Things shifted before that."

I thought of Julia's sister Sofie coming from Germany to help, and Julia's trip with Blandina. Julia was already suffering. She was far from home and family, married to a man who wasn't, per-haps, the easiest to love. And the physical and emotional demands of motherhood, the debilitating love and the mind-numbing tedium—they had, as Sarina saw it, already taken their toll.

Sarina looked for Abraham. "I'm still not seeing your great-great-grandfather, where is he in this?" She looked off into space again. "He's distant. I feel him being very distant." Julia showed Sarina two children: "I see a boy and a girl—she's showing me two." Sarina looked confused, and started conversing out loud again with the air to her right. Something had happened with a son, and Julia didn't get to be there for it. "OK, so was it that son? No, it wasn't that son. What do you want to tell me about these two?" The two kids, she said, were more like Julia's parents as far as taking care of her; they were more functional than she was. Julia needed tending. And her children—Anna, Delia, Bertha, Paul, Arthur, Julius, Teddy—suffered for it.

I asked Sarina if Julia had ever been chained to a radiator. "She's saying yes." But Julia told Sarina that she wasn't insane. "She knows that some people think that she went crazy, but she never lost her mind and she wants you to know that. Yeah, she shut down, but she always had her mind, and it's important for her to say that."

And why, I asked finally, is she a ghost? Sarina interrogated the air. Julia knew she was dead, Sarina said, and part of her had already left. But part of her stayed in the house, trying to correct something that had gone wrong. Julia couldn't leave the house; she simply couldn't. "I have never seen this before," Sarina said, one portion of a soul leaving, and one staying.

Julia's soul had fragmented, like her story. And I was beginning to wonder whether I'd be able to put it back together.

eleven

❧ THE UPPER TEN ❧

Bertha (right) and Delia Staab as teenagers.

The ghost version of Julia's life goes like this: In 1883, the baby died, and Julia shut herself in her room—her elegant, long-dreamed-of, newly realized room. She suffered and grieved, and her hair went white, and she never left the house again. This is the story in the ghost books and on the Internet; it is the story I wrote when I was twenty-four and living in New York.

But as I delved further into the records, I saw another story. Sarina was right: it isn't always one thing that undoes us. Life goes on after loss—even the loss of a child, as hard as that is for me to imagine. Julia had, in truth, been lucky in the health of her children. She still had six healthy children—seven including the disabled Paul—while many mothers of the era saw few infants grow to adulthood. Julia suffered immense sorrow, as grieving parents do. And she had suffered even before she lost the baby, from depression and a keen sense of displacement. But she didn't come completely undone—not at that point, anyway. She remained in the world.

Henriette died in July. But as I pored over old newspapers, looking for signs of Julia's further deterioration, I found that in fact, she seemed more active than she had been before. Her name began to appear more frequently in the newspapers. In early October, three months after Henriette's death, Julia and her family traveled east. "A. Staab, Esq, accompanied by Mrs. Staab and his three charming daughters, Misses Anna, Adele and Bertha, have left for New York City," reported the *Santa Fe Review*. "Mr. Staab will remain absent for about a month. Mrs. Staab will visit at the house of her brother-in-law"—Zadoc—"for several months"—to recover, perhaps?—"and the Misses Staab will commence a three-years-course at the seminary of Miss Froelich in New York City."

The girls had previously been educated at the Convent of Loretto, established by nuns the archbishop had imported to Santa Fe. While Abraham's Spiegelberg cousins established their own nonsectarian academy, Abraham and Julia preferred their daughters to be educated by nuns. The convent was a refined institution, with courses in such disciplines as reading, writing, grammar, arithmetic, geography, history, astronomy, "orthography" (spelling), natural philosophy, botany, and needlework. Classes were taught in French and Spanish "equally," and designed to develop the girls' "intellectual faculties" and train them in "the paths of virtue." The students said prayers in chapel each morning; Abraham and Julia seemed not to mind. It was not inexpensive—three hundred dollars a year, with additional fees for piano, guitar, drawing, Italian painting, and making artificial flowers. "The girls of this academy consider themselves the upper ten," Sister Blandina explained in her diary. "As far as money goes they are."

As far as money goes, Julia's children were in the upper one, the upper zero, the heavens-high Santa Fe ionosphere. There was no one in New Mexico much richer than Abraham, and it was expected that his children would have all the educational advantages of that echelon. The year before they left for school in New York, Julia acquired a governess for the girls. ("Wanted," she advertised in the *New York Herald*, "A Governess To Go to New Mexico and take charge of three young ladies; must bring good references and be conversant with the English, German, and French languages, and capable to teach music.") Julia stayed in New York to interview prospective candidates, keeping a room at the Rossmore Hotel, an ornate midtown establishment.

New York City was the heart of the nineteenth-century German Jewish diaspora—the Lehmans and the Goldmans, the Loebs and the Schiffs—Jews who had arrived, like Abraham, in the 1840s and 1850s and had peddled, traded, sold, and manufactured their way to dizzying wealth. It was the world to which Santa Fe's Jews moved when they made their fortunes—most of them, though not Abraham and his family—and to which the rest aspired: "a world," wrote Stephen Birmingham in his 1967 book *Our Crowd*, "of quietly ticking clocks, of the throb of private elevators, of slippered servants' feet . . . of sofas covered in silver satin." There were "heavily encrusted calling cards and invitations," balls and charities, "little boys in dark blue suits and fresh white gloves," girls in satin dresses, German governesses, English butlers, Irish maids, French chefs. Finger bowls were mandatory at the dinner table; rooms were littered with Dresden figurines and bronze cherubs and fringed lamps; damask, marble, pigeon's-blood velvet, pianos draped in Spanish shawls. It was a world of private ball-rooms and dinners for sixty, of a mass summer exodus to Adirondack camps and the New Jersey shore, and, every two years, "the ritual steamer-crossing to Europe." Abraham's brother Zadoc lived in that world. Though he was not as stratospherically rich as the Lehmans and the Loebs, he moved in their circles. The same rabbi from Temple Emanu-El on Fifth Avenue married his children; he joined the same clubs, attended galas at the same events: the Harmonie Club, the Purim Ball, the Hebrew Charity Fair.

New Mexico was a young territory, still in the process of cleaving itself from Mexico, and it was poor. The Santa Fe upper crust couldn't replicate the flamboyant wealth and grandeur of their East Coast brethren, nor could they hobnob in such rarefied Jewish circles—there weren't enough Jews in New Mexico to form a circle. But the Staabs made their best New Mexico approximation. They occupied the heart of Santa Fe, literally—with the huge storefront right on the Plaza and the big, towering house a few blocks away—and also figuratively. The New York Jews, wealthy though they were, kept to themselves; they socialized together and vacationed in the same places. The Jews in Santa Fe mixed with Gentile society in a way their New York brethren still could not or did not. In New Mexico, they were "Anglo," as long as they were white. The smaller distinctions that reigned in the East and in Germany—high WASP, low WASP, Irish, Italian, Jew—didn't seem to matter.

On Thursday afternoons, Julia and her daughters took callers. The butler, McCline—he was African American, not Irish—greeted visitors at the door and showed them into the parlor. There were many guests of all creeds: business folk and politicos, society wives and soldiers. They accepted each other as equals. As the army moved in and out of Santa Fe fighting the Indian Wars, the Staab girls found themselves in great demand among the bachelor officers stationed at Fort Marcy. They danced in the eye of the social whirl. There were teas, sewing circles, reading clubs, dances, balls, riding parties, champagne and oysters, boxes at the Albuquerque opera. The girls rode sidesaddle and carried gold-headed riding crops. The boys—Arthur, Julius, and Teddy, who also went east to prep schools—wore tennis whites and striped sweaters and looked every bit the nineteenth-century swells.

Julia was around for all of this. She went on family visits to New York twice a year and also took regular trips to Germany to visit her family. In Santa Fe, she visited friends and took callers. In his journals, the Swiss anthropologist Adolph Bandelier, who explored New Mexico's Indian cultures during the 1880s—combing villages and ruins for specks of bone and obsidian—wrote of receiving visits from Julia and her daughters, even from the mentally impaired oldest son Paul, and also of visiting them in their home. "Went to Staab's," he wrote, "and spent rather a lonesome hour there." Most of his visits seemed less lonesome, however: Julia and Bandelier's wife, Josephine, were friendly. Josephine visited Julia, and Julia also visited Josephine—Julia took herself out of the house and made some effort, it appears. They went on carriage rides together; attended the same parties, and spent a number of "pleasant" evenings together, Bandelier wrote. Julia had *friends*— which made me unaccountably happy. She may have been grieving and lost, but she wasn't always the recluse the ghost stories made her out to be, that I once made her out to be. She was out in the world still, visiting and being visited, attending cheerful parties. ("At night, party at Koch's until midnight," wrote Bandelier. "Mostly Jews. Pleasant.")

Julia didn't travel to New York with Abraham and the girls in 1885 when they returned to school for their third year, but she was very much in evidence in Santa Fe society—seen in December of that year at a housewarming party for a clothing merchant named Gerdes, a lively affair that began at 9:00 p.m, saw supper served at midnight, and featured dancing to the music of the Thirteenth US Infantry string band. She received callers at home on New Year's Eve, a regular tradition; she went to California with her two oldest daughters for three weeks in January 1887.

In February of that year, Abraham, Julia, and all three girls traveled to Colorado to celebrate the opening of the railroad line between Denver and Santa Fe. Abraham had, years earlier, transported a large group of New Mexico's territorial legislators to Denver in hope of bringing a direct line to Santa Fe—he had paid all the expenses for the trip and had given each legislator a top hat and gold-headed cane. The legislature soon issued the necessary bonds, and in 1887 the first train motored up the line from Santa Fe to Denver. It was composed of "seven coaches, six elegant chair cars and a Pullman sleeper," reported the Denver *Rocky Mountain News*, and it arrived in Denver at ten o'clock at night bearing Abraham, Julia, their daughters, and a hundred-plus other Santa Fe residents. There was a ball the next evening. "Mrs. Staab and the Misses Staab enjoyed the privileges of the dancing floor," the newspaper reported. Anna wore pink silk with white lace trimmings; Delia "an elaborate toilet of corn-colored satin"; both wore "some very fine diamonds." Julia was seen dancing, too, in "a handsome robe of ruby plush, white lace trimmings; diamond ornaments." She danced into the night. Perhaps she still mourned her lost child; perhaps she still suffered from various mental or physical ailments. But she was nonetheless out accompanying her husband and chaperoning her daughters. It seems she wasn't as damaged in the years after Henriette died as everyone believed.

Julia traveled to New York again in the summer of 1887, and she spent the winter of 1888 there with her daughters. In February 1889, Anna—Julia's oldest—married an Albuquerque Jewish merchant named Louis Ilfeld. Abraham added an extra room to the back of the house for the occasion, glassed on two sides to afford a view of Julia's gardens. "Precisely at 7," the *New Mexican* reported, "the bridal party was ushered into the double parlors, the bride's mother"—Julia—"leaning upon the arm of the groom and the bride following, escorted by her father." Anna wore a "magnificent robe of white faille Francais cut en train, with point lace and orange blossoms." A judge married them, not a rabbi; then the party proceeded to the Palace Hotel. At the entrance to the hotel "was one feature which deserves special attention," the newspaper gushed. "In small gas jets appeared, three feet long, the letters 'A. and L.' "—Anna and Louis, her groom. It was very resplendent, very American. Dancing commenced at nine, and the party ate dinner at midnight. Julia was there, out of her room, looking lovely—she wore a dress of red velvet. For a time at least, she found it possible to rise to these occasions.

John

◇◇

I ARRANGED TO MEET John, a ghost tour guide, at the obelisk that sits in the center of Santa Fe's Plaza. The obelisk was erected in 1868, two years after Julia arrived, to honor the territory's Civil War casualties as well as federal troops "fallen in the various battles with savage Indians in the territory of New Mexico." The word "savage" was chiseled away from the marble in 1973 by a hippie in a hardhat, then scratched back in and chiseled out again.

I wasn't sure I'd be able to find John among the homeless and skater kids milling around the Plaza, but it was quite obvious when I spotted him: Indiana Jones hat, leather vest, puffy-sleeved green tunic, cargo shorts, hiking boots, white tube socks. We shook hands and set off.

Our first stop was a Southwestern knickknack store on San Francisco Street, where, John told me, four basement spirits liked to throw clothes from the hangers. The clothing stayed put while we visited, however. Next we wandered through a saloon-themed restaurant—red velvet wallpaper, colored lamps, plank floors, a stamped tin ceiling, and a mechanical bull. The place had once been the site of a card room owned by the notorious madam Doña Tula. It was brimming, John said, with the ghosts of whores and gamblers. "The ghosts don't want to quit," he told me, "they're having so much fun." He gestured to the empty barroom with a dramatic flourish.

We moved on. Ghosts tend to congregate in places with a history of violence, John explained as we walked down Burro Alley, the short and once notorious city block where Julia had lived as a young bride. It was paved now, and a statue of a firewood-laden bronze burro moped on one corner. There were no storefronts or whorehouses or homes full of children along the alley anymore, just the cleanly plastered sides of buildings that fronted elsewhere.

Still, the place reeked of another time, and it occurred to me that spirits tend to be seen in places that seem as if their best days are behind them. John told me that ghosts were everywhere in this city with so many pasts piled together—Indian, Spanish, Anglo; New Spain, Old Mexico, New Mexico; the Wild West, the tourist Southwest, surges of conquering cultures washing through the desert arroyos like spring floods. It is often the dispossessed and defeated who come back as ghosts, I'd read somewhere, the historical voiceless, finding voice.

We walked toward the low-porticoed rectangle of the Palace of the Governors, where the Spanish had once ruled the Indians with iron swords. On the doorway, an inscription once welcomed visitors: VITA FUGIT SICUT UMBRA. Life flees like a shadow. It does now; it did then, and even more quickly. There were beatings on this spot, John said, and hangings. Indians were enslaved here and rose up against their captors; priests were slaughtered; the Spanish governor's head was tossed around the courtyard like a football. "It was the bloody tower of the Southwest," he said. Now it was full of angry and ancient spirits.

La Llorona was another of these angry spirits, John told me. An Indian beauty who married a conquistador and gave birth to two children, she despaired when her looks faded and her man ran off. This was an old folk tale common across the Hispanic Americas. In the version John told me, La Llorona threw her two children into the Santa Fe River, hit her head on a rock in her frenzy, and died—another overwhelmed mother. Now she wandered the banks of the river, which was not much of a river anymore but rather an intermittent stream channeled and deflected—a ghost river—looking for other children to drown. From the willows by the riverside, a bony arm would rise, a pale claw waiting to pluck wayward children and drag them into the water. She could sometimes be heard wailing by the river's banks. John gesticulated vaguely toward the concrete channel a few blocks away.

We moved on, lingering at the porch of a pretty Victorian art gallery whose previous resident, a prominent Realtor, had moved out very quickly after he encountered a black-cloaked, waxen-faced figure who smelled of rotting flesh and froze all the houseplants. This was an evil ghost, but they are rare—"Only five percent of spirits are really evil," John said. I wondered how he knew this, though I didn't ask. The Indiana Jones hat did seem to lend him some authority. No one knew, John went on, why this particular ghost was in this particular place. "Could there have been a hanging post there?" John asked. Ghost stories are full of rhetorical questions; there are never solid answers when we probe among the dead.

The cathedral's bells tolled the afternoon Mass, and John used the opportunity to tell the dubious story of Abraham and the cathedral, how Jewish money helped complete the upper half. "They say it's a cathedral on the bottom and a synagogue on the top," he joked. John had a lovely smile; there was a credulous sweetness to him. He pointed out the site of the archbishop's gardens, where my grandmother had imagined Julia and Lamy eating apricots and reciting French poetry. It is now a parking lot, all those cuttings and blooms and fancies interred in asphalt.

Then we came to La Posada. We sat in the reception area that wrapped Julia's old house, drinking lemon water and looking at its front door. A mezuzah adorned the doorframe, and I could see Abraham's gilt initials above the original entry—A.S.—as prominent as Yahweh's on the cathedral.

John shared some details from Julia's past, and just as Lynne had moved seamlessly between census records and dreams, John also had no difficulty transcending the line between fact and lore. He mentioned that Julia had watched from the top of the stairs while her youngest daughter entertained. That wasn't hard to believe at all. He explained that a child had died in the house; I knew this was true. But he also noted that children sometimes see a ghost baby teetering at the top of the steps, because Julia's son had fallen down the stairs and died. I hadn't heard this story. And he speculated that Julia might have died of a laudanum overdose—though there was also a chance, he said, that Abraham had killed her. "The sixty-six-thousand-dollar question," John opined, "is the nature of their marriage." Yes, it was.

We wandered through the front door into the narrow hallway between the bar and the library. We admired the opulent brasswork, and John, sleeves swaying with each gesture, reviewed the various spots where Julia's ghost had been seen: in the library, the bar, the Rose Room. He pointed upstairs in the direction of Julia's suite, and told me how the faucet in Julia's bathtub turns on and off of its own accord ("nineteenth-century ghosts are fascinated by plumbing fixtures"), and how the housekeeping staff never—"ever"—go upstairs alone, not since the day when a door slammed shut on a maid, the linens tossed themselves around, and the maid was unable to open the door to flee.

Julia's ghost, John told me, is rarely quiet for more than a month—but she also won't show up on demand. People who book the room seeking Julia rarely find her. I probably shouldn't expect to, either, he suggested diplomatically. It's the unsuspecting ones who have encounters. "Spirits are like colts," said John, "they're skittish."

twelve

❊ THE GREAT PACIFIC ❊

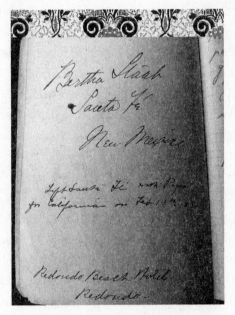

The first page of Bertha's diary.

In her family history, the dusty photocopy that I'd found in my great-grandfather's mountain home, Aunt Lizzie mentioned a diary that had belonged to her mother: my great-grandmother, Julia's third daughter, Bertha. It covered the years 1891 and 1892, when Bertha traveled to California with her father and then to Europe with both her parents. In Germany, Lizzie said, Bertha had written of a "dreadful accident" that befell Julia, but Lizzie's history didn't say what it was.

I decided, after reading Lizzie's book, that I had to find the diary. I wanted to know about the accident and Julia's trip to Europe, and more than that, I needed to see and feel the pages that had once documented Bertha's life, and perhaps her mother's as well. I needed to have this connection.

The diary had to be somewhere; Lizzie's children couldn't have just thrown it away. So I emailed Lizzie's daughter Nancy—Lizzie's only living child. She told me she had seen the diary only once, after Lizzie's death in 1980, and remembered the "fancy writing" inside. But her sister Judy had taken it. Judy had died in 2003, from a brain tumor, and Nancy didn't know what had happened to the diary after Judy's death.

Nancy suggested I try Judy's husband, Ron, who had relocated to southern New Mexico and, Nancy thought, may have brought the family copy of Hitler's *Mein Kampf* with him; perhaps he had the diary, too. I hadn't known there was a family copy of *Mein Kampf*, nor that it was something that Jewish families typically handed down through generations, like a prayer book or a menorah. I emailed Ron, who didn't know anything about the diary. He suggested I get in touch with his daughter Rhonda, who now lived at my great-grandfather's summer home and had taken possession of Judy's papers when she died.

Rhonda and I have always been close; she is eight years older than I, and as a child I idolized her. She is a free spirit with a thick mane of dark hair much like Julia's, though Rhonda accessorizes hers with daring turquoise jewelry, and she is full of opinions and sass. Rhonda and her brother were my only cousins to grow up in Santa Fe, and it was she who told me most of the ghost stories I had heard about Julia when I was a teenager. She was deeply connected to our family's history in New Mexico, so it surprised me, when I contacted her, to learn that she hadn't looked through Judy's papers in the nine years since her mother died. Perhaps this was her way of managing her grief—packing the ghosts away for a while.

When I called Rhonda to ask about the boxes of Judy's pictures and letters, she said they were piled up in a storage room and promised me she would go through them when she found a moment. A few months later, she and Nancy—up visiting from Albuquerque—examined the boxes, "& LO & BEHOLD," wrote Nancy in an email, they "picked up a little book that is the diary of Bertha!!!!!!! YEA." It was a small leather-bound notebook, Nancy said, three inches by six inches, and it contained writing so tiny that Nancy could hardly read it. A couple of weeks later I traveled down to New Mexico to see it myself. In it, I hoped, I could find some insight into Bertha's life, her time, her place, her family, and most of all, her mother. I longed to find Julia in those pages.

The dark-brown leather was a little scuffed, but the pages were still relatively undamaged, browned and crumbling at the corners but white on the inside. On paper lined in a thin, pale-blue grid, Bertha had tracked her days in a neat, penciled cursive. She had also tucked small aspen leaves into the pages to press and preserve them. The leaves were 120 years old when I first opened the diary, but they were no more brown or brittle than last year's dried flowers, their decline into dust suspended by care and happenstance.

I sat down to read the diary on the porch facing Hermit's Peak—the same spot where I first read Lizzie's remembrance of our family. I handled the book with the tips of my fingers, as if it were something exceedingly rare and precious, which it was—a direct line to Bertha and Julia, through Lizzie and Judy, Nancy and Rhonda, and now me. I couldn't believe my luck. Finally, the dead would speak—no more parsing rumors and reading between the lines.

On the first blank page, Bertha had written with a fountain pen. This was her only entry in pen—large, looping letters, with little blots where she lingered too long: "Bertha Staab, Santa Fé, New Mexico." In smaller letters below she wrote, "Left Santa Fé with Papa for California on Feb 14th, 1891." The diary continues in pencil, her voice more mundane than I had hoped and imagined it to be, detailing her activities both rousing and, more often than not, dull, beginning on the day Bertha and Abraham arrived in Los Angeles. Bertha had completed her three years of finishing school at Madame Froelich's and was living at home in Santa Fe with Abraham and Julia, waiting, I suppose, to marry. I couldn't ascertain at first the reason for the trip, but it was clear that this was an adventure for Bertha—her sisters had gone to California with Julia a few years earlier and it was now her turn to see the West Coast.

It had been an exhausting and jostling two-day train journey from Santa Fe. "Am tired when I write this at 11 o'clock at night— Have been sitting in lobby of Hotel all evening with a party of ladies and gentlemen who came with us on Pullman Sleeping Car—This morning at six got up to change cars at St. Bernardino," Bertha wrote.

The next day, Bertha took a carriage ride around Los Angeles. Like her mother, Bertha seemed to have a particular fondness for plants and flowers. "Saw palms, several varieties—the first palms I had ever seen and yet they looked as familiar as if I had seen them every day." She spied her first century plants—tall, hobnailed agave spires that didn't actually live a century, but could survive a good decade or three and flowered only once and died shortly thereafter. She saw pampas grass, and "the spruce tree in different varieties" clipped into hedges and arches and columns and gateways. She rode past orange groves and "smooth lawns with lines of calla lilies and great big bushes of geranium and climbing roses in full bloom at this season of the year"— winter! She found the homes that flanked those beautiful gardens less impressive. They looked "old, time-worn and faded," she wrote. Perhaps they were, compared with her father's nouveau-riche palace in Santa Fe.

Bertha was twenty-one, the same age Julia had been when she'd married and made her journey to America. In photographs, Bertha looks remarkably like her mother—tawny-skinned, with the same intense, close-together eyes; thick, silky, dark hair; half-moon cheekbones; and an ample bust. But she was a modern girl with expectations that were very different from her mother's at the same age. Julia had married a Jewish man from her small village in the old country and had subsumed her life into his; Bertha, nowhere near married, spent her time gallivanting with Santa Fe's elite. "This time to-night," wrote Bertha in Los Angeles, "might be at home at the Governor's reception, where I suppose at this minute they're having a goodly feast—Ah! for a word with _ _ _ _ ." She pined for a man whose name, even in the privacy of her diary, she wouldn't reveal. " 'I shall never see my darling any more,' " she continued, quoting a popular song about a slave whose sweetheart is sold away. Bertha was full of longing and self-consciousness and drama. She was bursting with the future, and I found her words sometimes agonizing to read. I can only imagine that my own journals from that age would seem equally cringeworthy to descendants riffling through them a century from now.

The day after her arrival, Bertha toured the Los Angeles basin. With her father and some friends they had met on the journey—a Mr. and Mrs. Crampton and a Mr. and Ms. K. Royce from Rutland, Vermont—she took the train to Redondo Beach to see "the Great Pacific for the first time."

Had an hour's trip fraught with no special interest except for the beautiful mountains—They loomed up in the far distance, their bases hidden by thick mists, their tops seeming to float in air and looking like immense white clouds.

Bertha clearly seemed to enjoy her time with her father; they had fun together. "Hired a three-seated rig for six persons. Managed to get seven in," she wrote. They drove toward Pasadena, through the wheat fields and past the palms and olive trees, past roses and heliotropes that grew high above their heads. "Met several 'Heathen Chinese,' " she wrote, "and accosted them but they vouchsafed us no answer." She took lunch at a grand hotel in Pasadena, drank champagne, tasted persimmon for the first time, bought a silver spoon, visited the ancient San Gabriel Mission church (which was "not as curious for us because had seen others just like it in New Mexico"), dropped in at a winery, tasted sweet "Angelica" wine, and then headed back to Los Angeles. "It poured down but we did not mind, fortified as we were with gossamers, umbrellas and thick coats." When winter rains socked in the next day, she fretted. "Been writing letters all morning," she wrote. Abraham played whist with the men while "ladies looked on," Bertha added. "Dull day."

As one would expect from such a personal document, Bertha's diary is much more about Bertha—where she went, what she did, how she felt, what she ate and drank—than Julia, who was home in Santa Fe, lingering just slightly offstage. And of course I hoped, as I scoured those fragile pages, that Bertha would shed some light on her mother's temperament and state of mind; that she might help me crack the code. But as I read, I found myself more and more engaged in Bertha's world—the world Abraham and Julia had created for her, and that now, as a young woman, she sought to make her own.

In the days that followed she shopped—buying another silver spoon—and socialized with other hotel guests. "Mrs. Schulte talked to us in her high cracked voice," Bertha wrote. It seemed that Bertha had a tendency to mock. She spoke about books with a fellow guest named Mr. Wright. "His pet expression is 'Well if you will' or 'If you do I'll eat my hand' which I recommended him to substitute for another expression." Her writing reminded me, painfully sometimes, of my own snobbish declarations at a similar age—the disdain of a young woman trying to decide who belonged in her tribe and who didn't.

Bertha took walks when the weather cleared, and she expanded her circle of acquaintances. It seemed not to matter whether they were Jewish or not—at least Bertha didn't mention it. Her judgments seemed to be rather more personal in nature. "Do not like Miss Smith's face," she wrote upon meeting a new crowd of guests.

> *Her nose is turned up—very suggestive of what she may be, but I do not know. Mr. B also forming one of their party is anxious to meet me thinks I have an intellectual face—First time I think anybody ever said that of me! The man must be of poor eye-sight.*

Bertha wrote often of men—"gentlemen," she always called them—and what they thought and said of her. She was a tad boy-crazy—more than a tad, actually—and as I read through the diary, I began to wonder whether Abraham hadn't dragged Bertha along on his trip to separate her from the fellow for whom she pined—an army officer, perhaps.

For Abraham, the journey had multiple goals. His health needed tending—he was, according to the Santa Fe papers, receiving "electric treatment." Bertha never mentioned what that treatment entailed, or what exactly ailed Abraham. But in addition to traveling for his health, it seems that Abraham had also journeyed to California in hopes of convincing the army to keep its troops headquartered at Fort Marcy in Santa Fe—the troops his business supplied. The War Department had recently announced that the Southwestern command was moving to California. So while Bertha shopped and socialized and explored, Abraham met with the military brass, hoping to persuade them to change their minds. "Met Col. Willard of the United States Army,"

Bertha wrote. "Said he did not think there were any prospects of having headquarters in Santa Fe." Indeed, a letter arriving by post soon informed Bertha that the army—including all the officers who had squired and admired her—was already preparing to leave Santa Fe. Bertha "felt blue" about this. She longed to be among the officers. Instead, she was in California engaging in unsatisfying flirtations with unsatisfying men. "Mr. Wright said I had a good forehead—Rats!"

In Los Angeles, Abraham's health did not rebound. They had planned to ride the cable car to downtown Los Angeles, but Abraham had a "weak spell" and sent Bertha with her friends. She saw a play, *Barrel of Money*, at the Los Angeles Theatre. "Could have been worse, but not much." She took walks, wrote letters, read books, toured another winery, visited another orange grove, bought another silver spoon—this one for her sister—visited an ostrich farm, collected shells at the beach, and on the evening of February 26, went to see a strange performance at the Pasadena Grand Opera House. It was a show that defied logic, presented by a performer named Annie Abbott—"an ordinary-looking woman of about 28," Bertha wrote, who did things that weren't ordinary at all.

thirteen

❦ MEN COULD NOT MOVE HER ❦

Annie Abbott, performing her levitation act.

Annie Abbott was a diminutive, graceful woman from Georgia "noted for miraculous strength," Bertha wrote in her diary, "which lies not in physical force but in some hidden power. Nobody knows what." Annie—her real name was Dixie Annie Jarratt Haygood—was a traveling performer who executed uncanny feats of strength and gravity. "She lifts a chair with six gentlemen upon it by applying the open palms of her hand to the chair," Bertha wrote. "They insert an egg between her hand and chair and she does not break it—wonderful." Audiences and scientists alike were baffled by Annie's power. It was thought to be magnetic, or perhaps electrical. "Some of them say I am possessed of the Devil," Annie wrote in a letter to a Georgia newspaper in 1893, "and others say I am another saint." She considered herself a Spiritualist—one who communes with powers from beyond.

It is convenient that Bertha ran across Annie Abbott, a wandering Victorian Spiritualist, and that I learned of their intersection as I considered the story of Julia, a wandering Victorian spirit. Because Julia's story fulfills a certain tradition: a woman, not a man, who appears late at night in a black or white high-necked gown, her hair piled elaborately on top of her head, and possessing an "aura of sadness," thanks to a life cut short, or things left undone and unexplained, or passions left unresolved. It is not a coincidence, I suspect, that this particular brand of ghost story—not the keening ghost in the white sheet, but the materialized human spirit in Victorian dress speaking lucidly from just beyond—emerged during Julia's lifetime. It was a time when Victorian spirits roamed the world.

Annie first performed her act in 1885 in her hometown of Milledgeville, Georgia. The next year, her husband died, leaving her with three children to support. She adopted the stage name Annie Abbott ("The Little Georgia Magnet") and took her act on the road, traveling the South and then the Northeast, captivating ever-larger crowds with feats of magnificent female power. Her act involved resisting male force; men could not move her. In New York, she faced off with the strongman Eugen Sandow, a brawny bodybuilding pioneer. "I put forth strength enough to lift eight men clear off the floor," he said, "yet I failed to shift this little fragile creature one inch from her position." Here and there a newspaper article would claim to expose the trickery behind her (the crafty use of leverage, illusion, and charm). No matter, the crowds kept coming. She traveled to Canada, California—where Bertha saw her—and Europe, where she packed theaters for weeks and performed before the German kaiser, the Austro-Hungarian emperor, the Russian tsar, Queen Victoria (whom she helped to locate a missing pearl), and the sultan of the Ottoman Empire (who watched the show, "but did not smile," because he believed her to be a witch).

Soon, other "Annie Abbotts"—imitators—sprang up, three or four or five of them in competition. Some dressed like little girls, others like Spanish flamenco dancers, others like proper Victorian ladies. But all of them were small, and all of them hefted men many times their weight, claiming powers that lay beyond our comprehension.

◇◇

Spiritualist performers could be found everywhere in the years when Julia lived in Santa Fe: young and old, men and women, in small towns and large cities. Their "talents" varied—some read minds, some talked to spirits from beyond, some, like Annie Abbott, exhibited feats of paranormal strength. Forebears to the psychics I met in my search for Julia's supernatural side—Misha with her tarot cards, Sarina with her lost son—they claimed to channel the powers of the dead. Julia might have consulted with such mediums in Santa Fe after she lost Henriette. It wouldn't have been unusual.

All of them—from Annie Abbott to the Stanley Hotel's ghost hunters—came out of a tradition born four years after Julia's birth. In March 1848, two teenage girls reported hearing a series of fearsome "rapping" sounds inside their bedroom in a rented house in the small town of Hydesville, New York. Maggie Fox was fourteen years old; her sister Kate was eleven. The rapping noises were so loud, wrote the girls' mother, Margaret, in an affidavit, that it shook the bedsteads and the chairs. Sometimes the raps sounded like knocking on walls, sometimes like the moving of furniture, sometimes like a person walking. The noises resumed the next night. "We heard footsteps in the pantry, and walking downstairs; we could not rest . . . ," Mrs. Fox wrote in an affidavit she signed a few days later. She declared that she was not, by nature, "a believer in haunted houses or supernatural appearances." But the noises were impossible for her to ignore, and she concluded that the house "must be haunted by some unhappy restless spirit."

On the third night—March 31, 1848—the family went to bed exhausted from the previous sleepless nights. "I had been so broken of my rest," wrote Mrs. Fox, "I was almost sick." She had just "lain down," when the noises commenced. Her daughters, sleeping in the room with her, heard them, too, and Kate, the younger one, began snapping her fingers in response. "Do as I do," Kate said to the unseen noisemaker. She clapped her hands: "The sound instantly followed her with the same number of raps," wrote Mrs. Fox. It stopped when she stopped. Kate repeated her clapping, varying the number, and the noisemaker followed suit each time. Mrs. Fox then gave the noisemaker a test. "I asked the noise to rap my different children's ages, successively." It did. It then informed the Foxes, through a sort of simplified Morse code of yes-or-no raps, that it was a spirit—a man, aged thirty-one, named Charles B. Rosma, who had been murdered in the east bedroom of the home five years earlier, cut through the throat with a butcher's knife and buried ten feet under the buttery below the house. Thus began a communication with "Mr. Splitfoot"—the girls' nickname for their visiting spirit—that first wowed the neighbors, then consumed the nation.

Hydesville, which no longer exists, was a hamlet about twenty miles from Rochester, in the heart of the "burned-over district," the region of upstate New York famously swept by wave after wave of evangelical religions in the 1830s and 1840s, from which emerged Mormonism, Christian Science, Millerism (which predicted Jesus' return on October, 22, 1844), and the Oneida Community (the silverware-making commune that encouraged sexual congress between postmenopausal women and teenage boys). The area was so heavily evangelized, it was said, that there was "no fuel left to burn." The Fox girls' story was one of those embers. They are credited as founders of Spiritualism, a movement that rested on the basic premise that humans could communicate with the dead.

As news spread of their communications with Mr. Splitfoot, the family was tortured by the ceaseless, sleepless rappings. The hair of the girls' unnerved mother went briskly white—Julia was not the only woman whose tresses succumbed to nineteenth-century traumata; it seemed to be a Victorian tradition. Only after Mr. Splitfoot informed the girls, via what must have been a rather elaborate series of raps, that they should "proclaim this truth to the world" did he let them sleep. The girls began to speak of their sensational communion with the spirits. Aided by a group of radical Quaker suffragist abolitionists and accompanied by their older sister Leah, Maggie and Kate traveled the state and then the country, visiting the homes of the wealthy and powerful and conducting public séances that were attended by such luminaries as William Lloyd Garrison, James Fenimore Cooper, Horace Greeley, and Sojourner Truth—sessions in which they conveyed messages from the dead on such weighty afterlife subjects as railway stocks, love affairs, and the existence of God.

It's not as if people didn't think about ghosts before the Fox sisters came along. In New Mexico, Spanish and Indian legends teemed with spirits: murderous mothers, vanquished Indians, and Kokopelli tricksters. Julia's village of Lügde, too, produced its share of ghosts: malicious wood spirits, grunting swine phantoms, a tax collector turned hellhound, and Sister Irmgard, a mad nun betrayed by a false lover, whose spirit wandered restlessly along a nearby stream, "shadowy and bloody," according to a book of local ghost tales. There are ghosts of legend and literature—Bloody Mary, the Headless Horseman, the Headless Nun, Banquo, Jesus, Hamlet's father.

Across the ages and across all cultures, people have claimed to hear from the dead. These tales remind us that there were people here before us and that others will take our places. But the dead of ancient legend did not, typically, communicate back and forth with the living quite so readily as they did in the Victorian era. Until the nineteenth century, ordinary people rarely boasted of speaking to specific dead relatives; their extrasensory perceptions weren't so finely grained. Victorians, on the other hand, held regular posthumous counsel with dear departed ones. The practice of mediumship—of the dead speaking through the living, as through psychics today—was something new.

In the years that followed the Fox girls' revelations, mediums emerged from every cabinet and closet. By 1853, there were more than thirty thousand in the United States alone. The Civil War, which left so many American families bereft, made communication with the dead only that much more appealing. The world was afire with talkative ghosts. Mary Todd Lincoln held séances in the White House in the hopes of talking to her dead son, reaching across the "very slight veil [that] separates us from the 'loved and lost' " (the president was in attendance); Queen Victoria is rumored to have tried to reconnect with the spirit of her beloved Prince Albert at Windsor Castle. Leo Tolstoy and Tsar Alexander II communed with mediums; Arthur Conan Doyle, whose fictional Sherlock Holmes was such a devotee of deduction, attended regular séances and wrote a book on the subject; Harriet Beecher Stowe reported that she had written *Uncle Tom's Cabin* under the guidance of spirits. Spiritualism infiltrated the fabric of everyday life. "It came upon them like a smallpox," the British logician Augustus De Morgan wrote in 1863. "And the land was spotted with mediums before the wise and prudent had had time to lodge the first half-dozen in a madhouse."

By the 1890s, when Annie Abbott traveled the world and Bertha Staab watched her perform, the Spiritualist movement was said to have more than eight million followers in the United States and Europe, drawn mostly from the upper and middle classes. All across America and throughout Europe, perfectly respectable citizens opened their homes to Fox-like mediums who fell into trances and produced ghostly "spirit guides," many of them clad in turbans and bearing Hindu-sounding names like Abdula Bay, Bien Boa, Uvani, Feda, Afid, and Nepenshis. The visiting ghosts, in turn, summoned other spirits—the Victorian dead—who indicated their presence by tilting and turning tables, levitating flower vases, writing on walls, clog dancing, exuding "ectoplasm" (the spirit ooze so memorably popularized in the Ghost-busters movies) from various orifices, and ultimately, if the spirits were so kind, "materializing" hands, or heads, or even full-fledged bodies clad in frilly high-necked dresses. Spiritualist mediums held trance lectures, teacup readings, and demonstrations of automatic writing. They sold aura photography, in which clients sat for photographs with their dead friends and relations. And they conducted vaudevillian displays of paranormal powers such as those claimed by Annie Abbott.

Not everyone accepted these claims without scrutiny. The most popular mediums found themselves examined extensively by doctors and researchers. Scientists at the time didn't shy away from such pursuits; such intellectual luminaries as Thomas Edison, Marie Curie, the physicist Sir William Fletcher Barrett, and the British chemist William Crookes all dabbled in paranormal research at one time or another. In America, the pioneering psychologist and philosopher William James founded the American Society for Psychical Research (ASPR), a spin-off of a British group aiming to prove—or refute—the validity of paranormal phenomena by subjecting claims of table tipping and telepathy to the withering light of scientific method. All one needed to prove that ghosts could exist in general, James explained in a famous 1890 lecture, was to prove that one ghost existed in specific. "To upset the conclusion that all crows are black," he said, "there is no need to seek demonstration that no crows are black; it is sufficient to produce one white crow; a single one is sufficient." One real ghost would be enough.

In the hopes of finding that one real ghost, James's paranormal researchers devised all manner of experiments and contraptions— objective measures that might explain or debunk the phenomena they witnessed. Commissions formed; reports were issued; leading scientific and academic lights weighed in on the subject of ghosts. Lots of tom-foolery was detected: mediums' cabinets with hidden compartments; psychics levitating stools with their feet from the safety of billowy skirts; ectoplasm that, upon further scrutiny, proved to be cheesecloth, or toothpaste, or gelatin and egg whites. But there were also investiga-tions that defied explanation.

The Fox sisters stumped the scientists. William Crookes, the prom-inent British chemist who worked with the British Society for Psychical Research, examined Kate Fox and found her powers—spirit writing, lights, teleportation, materialized hands, and her famous rapping—to be formidable and inexplicable. "With mediums, generally it is neces-sary to sit for a formal *séance* before anything is heard; but in the case of Miss Fox it seems only necessary for her to place her hand on any substance for loud thuds to be heard in it, like a triple pulsation, some-times loud enough to be heard several rooms off," he wrote. He heard the thuds when her hands and feet were held; he heard them when she was "suspended in a swing from the ceiling," when she was enclosed in a wire cage, "when she had fallen fainting on a sofa." He felt the raps on his shoulder and under his hands. "I have tested them in every way that I could devise, until there has been no escape from the conviction that they were true objective occurrences not produced by trickery or mechanical means," he concluded.

A scientist from the ASPR examined Annie Abbott, too. She stood on one foot and held a pool stick horizontally in front of her in her open hand while four men tried to force it down and upset her balance. After two hours of this, reported Dr. Lewis G. Pedigo, an ASPR researcher, Annie had defeated "the strength of a dozen vigorous, athletic men," but was "perfectly fresh and free from fatigue." In a twenty-six-page study of her powers, he concluded that her strength came not from "a new force, but a rather unfamiliar manifestation of an old one—viz: nerve force."

What is remarkable, from our vantage point today—from my vantage point, exploring this unfamiliar paranormal culture—is how few people thought these investigators crazy. Much of what looked convincing back then seems flagrantly fake today: obviously doctored photographs, women with ectoplasmic arms made of sheep intestines and cheesecloth adhered to their torsos. But at the time, all sorts of new, invisible, and inexplicable forces were capturing sounds and voices and images from the ether. Electricity, for one. Telegraphs— some mediums took dictation from spirits in Morse code—telephones, photographs, X-rays. These technologies were no easier to understand and believe, for the vast majority of Americans and Europeans, than unseen departed souls communicating from beyond—and yet they worked. Modern life required a degree of credulity.

My sessions with Sarina and Misha made me uncomfortable. It is not considered rational, in this century, to schedule meetings with the dead, and I worried that my friends and colleagues would think me softheaded. But in Julia's time, perfectly respectable people—scientists, journalists, presidents, scholars—conversed with the departed, and weren't afraid to admit it. They listened to young women barely Bertha's age, and imagined that they had something important to say. Women—whether they were working-class girls like the Foxes, or wealthy women like Bertha and Julia—had little control over their worlds. They served their parents, then their husbands and their children. But standing in front of a

Spiritualist audience, Maggie and Kate Fox and Annie Abbott commanded esteem. These women had a voice. They had a number of voices,
actually, many with odd Hindu accents.

◇◇

Spiritualism gave these nineteenth-century women a source of power
and respect they could not otherwise expect to hold. But this public
life was neither quiet nor easy. Annie Abbott traveled the world prov-
ing her strength with a succession of different husbands and lovers
(seven, total) who inflicted a succession of humiliations: desertion,
fraud, embezzlement, bigamy, infidelity, domestic battery, perhaps
even infanticide. One husband took off with Annie's money; another
left her and found a new Annie Abbott to shill for. Annie abandoned
a twin son and daughter. She suffered "noises in her head," debilitating migraines, and mysterious blackouts, and as her fame ebbed, she
was reduced to selling trinkets from her European tours. She even
sold her sixteen-year-old daughter, Maud, into marriage with a man
three times her age. Maud ran away and Annie never saw her again.

Annie probably had a morphine problem. She accused many people
of robbing her—mainly of jewels; they were always stealing jewels.
When she accused her oldest son of stealing from her, he had her arrested on a charge of lunacy. She was accused of kidnapping young
children and of nonpayment of taxes. She died in 1915 at the age of fifty-
five, indigent and alone, remembered by few. "It is one of the ironies of
fate," wrote a local newspaper, "that such a woman should die and even
her own generation forget all about her." Her grave, they say, is cursed.
Spiritualism gave Annie Abbott a voice, but it came at a price.

It didn't end well for the Fox girls, either. They had both, by their
mid-thirties, developed severe drinking problems, and they were able
to navigate public appearances only with the help of canny handlers. In
1888, the two sisters convened an audience at the New York Academy

of Music and, in front of two thousand onlookers, demonstrated that by cracking their toe joints, they could produce, on demand, rapping noises that could be easily heard throughout the large theater. They both signed a lengthy explanation of their frauds in the *New York World*, a confession for which they were paid fifteen hundred dollars. Maggie explained that they had dreamed up the noises to scare their mother, and that the sisters then had found themselves trapped in a lie and learned to profit from it. Maggie denounced Spiritualism as "an absolute falsehood from beginning to end, as the flimsiest of superstitions, the most wicked blasphemy known to the world"; Kate issued a similarly scathing statement. Maggie retracted her words a year later, but it did her no good; the magic was gone from the act. Within five years both sisters had died in poverty, shunned by their friends and former supporters.

That might have been the end of Spiritualism, except it wasn't. Believers remained convinced that the Fox girls had spoken to spirits and that the women's confessions, swayed by dire need, were fraudulent. In 1904, nearly six decades after the first queer knockings in Hydesville, the *Boston Journal* reported that a basement wall had fallen in the Foxes' old house. Investigators found "an almost entire human skeleton between the earth and crumbling cellar walls," along with a peddler's tin box. The finding, said the paper, cleared the Fox sisters "from the only shadow of doubt held concerning their sincerity in the discovery of spirit communication."

Though of course it did no such thing. When you traffic in spirits, doubt casts a long shadow. People will believe exactly what they want. And yet we keep trying to exhume those skeletons—I kept trying to understand my family's history, stringing together clues real and imagined, hard facts and softer "spirit communication," hoping to find a story that felt something like truth. We are all mediums, who try to connect to the past. Like Annie Abbott and the Fox Sisters—like Julia perhaps—we hear voices.

The Reverend Conklin

◇◇

IN CASSADAGA, FLORIDA, THERE is a picture of the Fox girls' childhood home. The framed black-and-white photograph is nailed to the wall of a wooden meetinghouse—a spare, camp-style building with a pitched roof and folding chairs. I went there on a breezy day in March with my mother-in-law, who had recently lost a brother. She wanted to learn of his fate; I wanted to learn of Julia's.

Cassadaga is a Spiritualist colony—the Southern Cassadaga Spiritualist Meeting—one of the few remaining, a modest remnant of the much larger nineteenth-century movement. It is sometimes referred to as the "psychic center of the world." It is about an hour from Orlando, where my in-laws live. The camp was chartered in 1894 by a tubercular upstate New Yorker named George Colby whose spirit guide, Seneca, led him from a séance in Iowa to a ferry landing on Florida's Saint John's River. He hacked his way through thick palmetto and pine forest over seven hills to a spot overlooking "a chain of silvery lakes." There he was cured of his tuberculosis and established a Spiritualist outpost.

These days, the psychic center of the world consists of a tight grid of sleepy, sandy streets with a scattering of clapboard buildings, a tin-roofed camp headquarters, a gas station, a post office, and a two-story stucco hotel that offers such services as past life reflection, chakra balancing, auric repair, crystal healing, and, in case you need it after all that psychic tumult, "hair design." After we arrived, my mother-in-law, Toni, wandered over to the hotel to see if she could scare up a reading. Toni has a high voice and near-buzz-cut graying hair, and is as large in pluck and persona as she is small in stature. She suffers from none of the sheepishness about communing with the dead that I do—she encounters dead relatives everywhere. I was sure she would learn lots of intriguing news about her departed; I was less convinced of my own success. I headed to a large tin-roofed camp building to meet with the Reverend Judy Cooper, the camp's media liaison.

We liaised in a large meeting room behind the camp store, sitting down in folding metal chairs to chat. The Reverend Judy looked to be around sixty years old, enveloped in a pouf of blond hair and accessorized with glittery nail tips, pearl earrings, and a pink sweater with pearl buttons. I told her about my search for Julia and explained that I was here to understand, if I could, the things that eluded me in the books and newspapers and genealogies. Was Julia truly a ghost? And why was it that she was so unsettled?

Reverend Judy asked me if I had tried to contact Julia on my own, and I told her that I didn't know how. She suggested that I could, through meditation, connect more deeply with my spiritual side. My "gut," she said, was my door to the spirit world. "You get gut feelings all the time but then try to rationalize them," she explained. "Your gut feeling is a little muscle. The more you trust it, the stronger it will get."

My gut was weak from a lifetime of neglect. I didn't trust it at all. But I also didn't rule out its existence. Once many years before, a friend and I, terrifically bored in the rain on a camping trip, played a game in which we tried to guess the next card in a deck. I guessed wrong every time, until about halfway through the deck, when I suddenly and absolutely knew that the next card would be the four of clubs. And it was. That's why I was here.

After Judy and I finished chatting, I found my mother-in-law—back from communing with her brother—and joined a group of seven or eight visitors for a tour. Our guide was named Jeri. She was a Reiki master who was in training to be a medium. It was a rigorous curriculum that involved four to seven years of work to develop her "gifts of clair": clairvoyance (seeing things); clairaudience (hearing voices); clairsentience (feeling and knowing, the gut feeling Reverend Judy Cooper described); and clairgustience (tasting and smelling, cigars, roses, perfume, fish). Jeri had been clairsentient all her life, she explained, but she had thought she would never be clairvoyant, because she wore glasses.

Not so. Thanks to her rigorous training at Cassadaga, Jeri was gaining new gifts. Recently, in the middle of the night, she had seen a "white, scintillating" energy emanating from crystals on the dresser next to her bed.

We strolled down Spiritualist Street, past a tidy row of tin-roofed clapboard houses. Just past the intersection with Mediumship Way, we stopped at a charming cottage constructed a century before from the Sears and Roebuck catalog. The house, Jeri said, was often visited by the spirit of an eight-year-old girl named "Nietzsche." If you wake up and find pennies, Jeri said, you know that Nietzsche has been there.

Jeri pointed out other homes, other spirits—ghosts of Julia's vintage in Victorian clothing. A number of homes included odd second-story doors that opened up to nothing: no balcony, no fire escape. They were "spirit doors," human-sized exits from the pitched-roof upper floors that once housed séance rooms. In the colony's early days, the occupants believed that ectoplasm—spirit ooze—needed a quick way of escaping the room; hence the spirit doors. "Since then we've realized that spirits can go out through walls," Jeri explained, "and we don't need that."

We stopped at a sunken circle of head-high dog fennel known

as Spirit Pond—it was formed from the overflow of three lakes that rarely overflowed anymore, and it was dry. If we took a photo between two palm trees next to what had once been the edge of the water, Jeri said, we might see orbs; green ones indicated a human spirit. Three closely clustered palms nearby, she said, were called the Elevator Trees. "Put your hands on the trees and just relax," she said, "and you can feel the energy coming up," like a "spirit elevator." She asked if we wanted to try it out. A lady in pink went first. She stood facing us between the palms expectantly, and waited. She waited some more, then walked away.

I went next. I placed my hands on the rough bark, closed my eyes, and felt nothing. I wondered if perhaps the soles of my shoes were too thick. My mother-in-law went after me. And the moment—the very moment—that she put her hands on the trees, she shivered and made a surprised, ticklish sound. "Whoo! That is really—ooh!" The woman in pink looked disappointed; I imagine that I did, too. "What are you supposed to feel?" the woman asked. Toni said she felt the energy rising upward through her feet, making her whole body numb. Jeri explained that my mother-in-law had clairsentient gifts.

I, clearly, did not.

Next I met with the Reverend Ed Conklin, a renowned Spiritualist minister whose great-great-grandfather had given more than thirty private readings to Abraham Lincoln, including one in which the spirits dictated the text of the Emancipation Proclamation. Reverend Conklin lived in a white clapboard house with peacock-blue shutters and a souped-up black Honda coupe parked on the Bahia grass in front. He was in his seventies with a white beard and slightly sunken eyes. In his plaid shirt and down vest, he looked as if he should be splitting wood in Maine instead of reading fortunes in central Florida.

We walked through his cluttered living room to a sitting room in the back, where the Reverend Conklin settled into an armchair surround by statuettes—buddhas, Indian goddesses— and explained how the session would work. "I'll relax, the spirits will come to me. Hopefully, they'll come. About half the time they do. They can give me a word, a sentence, a rhyme, a letter, sometimes a name." He got a look in his eyes, as if he were blind. "Now, I think you're much too young for Dad to be in spirit," he began. "If he's living they're referring to him. Does he have kind of a big laugh?"

My father's laugh is almost silent, high-pitched. "Just a regular laugh," I said.

"What about his father or mother, your grandparents. Did one of them have a hearty laugh?"

G-pop, my father's father—Julia's grandson—did have a distinctive, chuckley sort of a "heh-heh." I mentioned it to the Reverend.

"OK, I feel that's him coming through," the Reverend Conklin told me. "Did he get to the age where he had false teeth?"

Yes, I told the Reverend Conklin, he did use dentures, though I didn't know it until the very end of his life. He died at ninety-seven, after undergoing two hip replacements within three months. I was there for the second surgery. We didn't know if he would survive it—he was so frail, so outrageously old. His skin hung from him, his face was skeletal, his spine twisted. But he was as sharp as ever. G-pop was a man who knew how to live, how to grasp every passing minute. I loved this about him. We held hands and talked about the future.

Shortly before they rolled G-pop into the operating room, he took out his teeth, which I'd never known were false. Toothless, he looked even older, if that was possible. They rolled him out of the room while he held the hand of his wife, Marge. He survived the surgery but died a few months later.

The reverend sat still, gazing above my head. A stub-tailed white cat circled his legs. I wondered if he was going to get to Julia soon.

But he was still focused on G-pop. Now, the Reverend Conklin told me, my step-grandmother Marge was coming to him. She told him that I had a yellow energy. She also said that I might get married, and that I might soon become a physician's assistant. My hour was ticking down. Where was Julia? I tried to listen to my gut, as Reverend Judy had counseled—but at the moment the only thing it was telling me was that Reverend Conklin's gut didn't seem to know anything at all about me or my ghosts. Finally, I told him that I was looking for a relative named Julia.

He offered a small, reflexive smile. "Did you know her?" he asked.

Oh Lord.

I didn't, I told him.

He nodded slowly. "I think she lived to a good old age," he said. I didn't bother to correct him. I asked instead how she died.

Julia showed him a lot of rich foods—pies, roasts. He thought she might have suffered from arteriosclerosis.

He felt she was strong-willed. "I like her poise," he said.

She had a blue energy about her, he said, and she also made him feel something in his umbilical cord—a family connection. She cared about her family, he explained. "I feel like she was a little conservative," he went on, "like if a man attempted to flirt with her, she would say he's fresh."

But he didn't feel Julia's presence strongly. She was detached, as if she had been away from the world of the living for a long time, and had begun to lose interest.

Still, she was interested in what I was doing. And her interest in my project, he said, was about my journey, not hers. I was going to grow from this. I was going to learn something.

fourteen

❧ PROPER GIRL ❧

Bertha Staab as a young woman.

Reverend Conklin had had more misses than hits, but he was right about one thing: Julia was prudish. Bertha felt so, anyway. While in Los Angeles, she mused about it in her diary. A few days after the Annie Abbott performance, Bertha had been sitting in her hotel room writing letters when, she wrote, "I opened the window and looked out upon the porch nearby." There, standing on a ladder right beside her

window, was a certain Mr. Parkinson—a young fellow who had recently checked into the hotel. "I had not noticed a ladder coming right
into the window; Chinaman had been washing windows. . . ." Bertha's
first reaction was to draw her head back; her second was to lean back
out and chat with her visitor. "Mr. Parkinson sat on the ladder and [I]
sat at the window. Not very aesthetic to have a young gentleman looking in at a young girl's window, but of course we, like idiots, did not
see into that till afterwards. . . . I know what mamma would have said!"

She didn't actually say what Mamma—Julia—would have said,
though I got the sense from Bertha's diary that it probably would have
been expressed in German—sternly, and with disapproval. "How differently we are brought up from most American girls," Bertha wrote.
"Miss Keck"—another hotel guest—"calls her mother Old Curiousity! And sometimes Beauty etc. Respectful! Yet she is a very nice
girl. She simply does not think anything of this way of talking to her
mother—not her fault—but mistake in bringing up."

A clue, here: Julia was a creature of the Old World, her ideas of propriety formed in Germany and hardly altered by twenty-five years in
America. Even after so many years, she didn't seem much interested in
bending herself to fit. Blandina's diary and the long visits to Germany
suggest that she had never quite been able to; as she grew older, she
would only have become less supple. Bertha would never have called
Julia "Old Curiousity," though she might have found her mother a
curiosity—an old-fashioned woman in a newfangled country. Bertha
was a proper girl, of course, but she might have liked to expand the
definition of what a girl could do. "I think in some cases," she wrote,
"we ought to have more freedom for mistakes in accepting invitations.
I think we ought to be able to do as we like, except if there are grave
reasons to the contrary." The ladder-side chat, it seems, was as close as
she got to rebellion in Los Angeles.

Bertha and Abraham left the city soon after. On Abraham's birthday, February 27, 1891, he called on a General McCook, who delivered bad news about the Santa Fe army base. "Said he had been in favor of keeping post in Santa Fe," Bertha wrote, "but had been asked by Gen Schofield whether it was a military necessity, and had replied no." The next day, Abraham's health again took a turn for the worse—his spells seemed to coincide with bad news from the US Army. Maybe his nerves weren't as steely as they seemed. The prospect of losing the patronage of the army fazed him even more, it seems, than had Billy the Kid and his "broncho maneuvers."

Or perhaps it was just something going around—Bertha, too, was feeling unwell. "I have a vile cold," she wrote. "Feel dreadfully about everything, about myself, about Santa Fe about—I don't know what not. I am thinking how dull it will be in Santa Fe with all the officers away—oh!"

On the advice of a friend, Abraham traveled to see a Dr. Millard in Redondo Beach, who counseled him to stay there under his care for a month or two. Abraham agreed, and he and Bertha packed for the move to Redondo. "Lost my sapphire ring," Bertha wrote, "looked everywhere and could not find it." She worried that Abraham would be angry. But while he may have been a tough businessman—and if the ghost stories are right, a domineering husband—he could be an indulgent father. "Papa is so good; instead of scolding me as I deserved, said, 'Well! If you don't find it I'll buy you another one.' " This Abraham—the doting father, the winning travel companion, the anxious army supplicant—seemed more human than the man I'd encountered through the psychics and the ghost stories. I was starting rather to like him, and it felt a relief to think that he might not be as bad as the ghost stories made him out to be.

While Abraham convalesced, Bertha struggled to entertain herself. "Tried a hand at billiards," she wrote, "and made a fool of myself." Redondo was considerably less lively than Los Angeles. "No gentlemen to dance with," she lamented. She started to read a new novel—*Daniel Deronda*, George Eliot's long, weird story about an English gentleman who rescues a drowning "Jewess" and finds out that he is also a Jew. After learning that Bertha had read it, I did, too, hoping in some way to see the world as she saw it. It's a tale of the drama and expectation of youth, peppered with elaborate proto-Zionist digressions. But Bertha had nothing but time. She consumed it far more quickly than I did.

When a new batch of young men finally arrived at the hotel, Bertha found that she had competition for their attention. The railroad sleeping-car tycoon George Pullman, who had invested in land with Abraham in northeastern New Mexico and was a friend, was staying at the same hotel in Redondo with his daughters. "Get mad," Bertha wrote, "because there's music here every night and nobody to dance with.—generally two Miss Pullmans have the floor—dance very well—elder has hard face. Younger more pleasant face and animated." The Miss Pullmans, she noted, dressed beautifully. "They have beaux coming out to visit them all the time. Probably invite them."

The Staab family's primmer code of behavior didn't allow Bertha to invite young men, as the Miss Pullmans did. But it was a testament to how far the Staabs had come that Bertha was writing about them at all. She had, until age twelve, lived with her Jewish-immigrant parents and six siblings in a dirt home on Burro Alley. Now she was jockeying with the Miss Pullmans for dance partners on the California coast.

Between unproductive forays to the dance floor, Bertha wrote and received lots of letters. She corresponded regularly with a "Col. Frost," and with her brothers Teddy and Julius, who were at boarding school in Philadelphia. Teddy got the measles; Julius won a silver cup for best athlete. Bertha played tennis with her friend Miss Keck—she lost four games to three—and finally, at long last, found some young men to dance with: a rather haughty Englishman and a very nice "Scotchman" who danced "horribly—round and round until you get 'giddy' as he expresses it." But after a few nights, the Brits, too, felt the pull of the Miss Pullmans. "Last night in ball-room the Scotch and English thingamagigs danced with the Misses Pullman—did not come near us. Do not consider that very gentlemanly behavior especially as they knew us first. . . . But so goes the world!" When the two men left a week later, however, Bertha confessed to her diary that she missed them "very much."

Abraham wasn't very much in evidence in Bertha's diary during this period in Redondo—recovering from his ailment, perhaps. Nor did Bertha mention Julia, though I found her in every line Bertha wrote—in how Bertha saw and responded to the world around her; the world her mother had helped to shape. On Julia's birthday—March 11, her forty-sixth—Bertha and Abraham sent a telegraph, and later, a bouquet of wild onions and protea flowers.

Finally, one very long and kabbalistic George Eliot novel, dozens of letters, and too many empty dance cards later, Bertha reported that Abraham was feeling better. They moved on to the Hotel del Coronado in San Diego for a few days, visiting with friends from home and catching up on the Santa Fe gossip, "which could have been compressed into a nut-shell," Bertha wrote—while Abraham sought, one last time, to convince the army to keep the post in Santa Fe.

It was to no avail, the troops would go. "What will Santa Fe do?" Bertha lamented. And what would Bertha do, with all the officers, and presumably her suitors, hauled off to another post? She suffered from enervating, excruciating boredom, and also a touch of despair. "I'm so fatigued," she wrote, "why cannot I have the person I love"—and whose identity she never revealed. "Why am I poky and horrid?" And then, at long last, they took the train home, where the soldiers were leaving, and Julia was ailing.

fifteen

❊ REGION OF INSANITY ❊

BUCHANAN'S SYSTEM OF ANTHROPOLOGY.

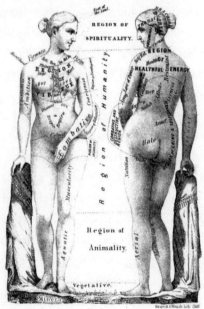

OUTLINES OF SARCOGNOMY.

Nineteenth-century doctors associated the womb with mental illness.

Bertha didn't stay in Santa Fe long. A month later, on May 10, 1891, Abraham, Julia, Bertha, sister Delia, and brother Paul—the disabled eldest son—boarded a train bound for Philadelphia.

This time, the trip was for Julia's health. She seemed to have recovered, for a time, from the spell of depression that Sister Blandina tended to, and also from the loss of Henriette. She had, for a few years, at least, rejoined the world. But now, Julia wasn't at all well. "In Chicago mamma was sick for 3 days and we were all frightened, especially about having her travel in her poor condition," Bertha wrote.

Was it Julia's sadness resurfacing? Or was she suffering now from something more physical in nature? Bertha's diary notes only that the family hoped that a visit to Germany could improve Julia's condition. Once again, Germany was where she went to heal.

When Julia had recovered enough to move from Chicago, the family forged on to Philadelphia, where they visited Arthur and Julius and Teddy, who were in prep school there. The family then spent another ten days in New York, shopping for dresses and silver spoons, visiting with their cousins, and packing. They sailed for Europe on the twenty-sixth of May. "First two days of our journey were pretty fair," Bertha wrote.

> *Delia was sick, the rest well. The third day was horribly rocky, the next also; mamma was dreadfully sick! Delia and I also—Papa during one night. . . . Met several pleasant people but did not make friends—Max Foster from Berlin, glove man was the nicest gentleman we met. The rest of the first cabin passengers were horribly poky and stiff.*

They docked in Liverpool, then Bremen, and set off by train to Hanover. Julia's father had died some years before, and her mother, Henriette—Jette, they called her—and Julia's remaining unmarried sisters had relocated from Lügde to Hanover. In fact, all of Julia's siblings had left Lügde. The brothers, Ben and Bernhard, emigrated to America. The married sisters—Adelheid, Sofie, Auguste, Bernhardine, Amalie, and Emilie—moved with their husbands to the nearby cities of Paderborn or Hanover. Their families owned mills, department stores, and legal firms, and their children went to elite Catholic gymnasiums and Swiss finishing schools. Since Germany's unification in 1871, Jews had enjoyed the rights accorded to all of the country's citizens. Julia's sisters were city Jews now, living in grand houses in the hearts of German cities. Their homes boasted music rooms, Renaissance art and modern masters, and libraries lined floor to ceiling with leather-bound books.

There was no reason to visit Lügde anymore, so Julia and family went straight to a hotel in Hanover. The European hotels did not, in Bertha's esteem, live up to the grandeur of those she had visited in New York and California. "No carpets in hotels in Europe, no running water, no gas—but feather beds and candlesticks and little mats in front of wash stand and bed," she wrote.

After a few days' visiting, the family split up. Abraham set off for Carlsbad, the famous spa and mineral springs in what is now the Czech Republic. The spa was a regular stop on the European vacation circuit, frequented by the ill and healthy alike. Abraham went to Carlsbad regularly—he had, the *Denver Post* reported, "been able, for a long time past, to take his rheumatism annually to Carlsbad and leave it there." Abraham's brother Zadoc had also been a frequent visitor; in fact he had died there, in 1884, while being treated for a liver "affection." Most visitors, however, had more encouraging outcomes. Spas were where the well-heeled went to rest and restore themselves. Which Abraham did, promptly, upon the family's arrival in Europe, while Bertha stayed in Hanover "to take gymnastics," and Julia, accompanied by Bertha's older sister Delia, went to Berlin. There, Bertha wrote, "mamma consulted Olshausen."

Robert von Olshausen, I learned from a quick Internet search, was a German gynecologist and surgical innovator. While neither Bertha nor Sister Blandina elaborated on Julia's particular ills, here, in Bertha's diary, was a clue: Olshausen's department at the Berlin University *Frauenklinik* was the leading women's surgery hospital in Germany, and Olshausen had been a pioneer in vaginal hysterectomy. Julia, then, suffered "female problems," as the Victorians put it—a physical malady, related, perhaps, to her many pregnancies and miscarriages; or perhaps her approaching "change of life" (menopause) had set off another cascade of hormones that left her feeling unbalanced and unwell. Whatever the cause, she was considering having a surgery to remedy the problem.

There were good reasons that she traveled all the way to Berlin to seek medical help: the doctors there were the best in the world. Germany had, by the time Julia fell ill, developed rigorous medical programs that involved years of university study, experimental and clinical work based on scientific method, and surgery using new antiseptic techniques. Physicians in the United States, by contrast, were marginally educated, poorly compensated, ill-trained journeymen peddling purgatives and patent medicines, advising on disease-causing vapors and "exhalations," and delivering what the historian David Dary called the "dreaded triad of heroic medicine"—bloodletting, purging, and emetics—to expel and rebalance bad "humors" in the body. American doctors had, since the Civil War, begun to discover and embrace germ theory, anatomy, and empirical inquiry, but much more slowly than in Germany. Things took even longer on the frontier.

There were few doctors in Santa Fe when Julia lived there. She saw two that I know of: Dr. Symington, the physician who had accompanied her and Sister Blandina to Trinidad in 1877, and W. S. Harroun, who would remain Julia's doctor at the end of her life. I had found Dr. Harroun's papers at the history museum in Santa Fe on one of my early trips to the archives: a folder of small notebooks, their sides burnt with age like lightly browned marshmallows. I leafed through them hoping for notes on Julia's ailments and the reasons she traveled to Germany to consult Olshausen. Harroun's ledgers noted such expenses as rent, heating wood, a saddle, a bay gelding, a bathtub, hay, shoes, horseshoes, club dues, and pants. He also listed payments from patients. When they paid their bills, he marked the entry with a smiley face, along with the ailment for which they'd consulted him. He frequently treated "pistol shot wounds." One bullet hole, he noted, was large enough to admit his little finger.

The notebooks told me plenty about what it was like to be a doctor in nineteenth-century Santa Fe—there were no receptionists, or insurance companies, or stainless-steel instruments—but nothing about what it was like to be Julia's particular doctor. None of Dr. Harroun's entries, sadly, mentioned visits with Julia. I imagined the doctor staying late in his office, scribbling his notes by gaslight. A messenger might come, and Dr. Harroun would grab a saddlebag of instruments, mount his bay gelding, and ride across the Plaza to see Julia, ailing again.

Even if Dr. Harroun's papers had mentioned Julia's complaints, they mightn't have told us much: doctors then treated *conditions*— fevers, catarrhs, dropsies—rather than specific diseases. In the journals of the Staabs' friend Adolph Bandelier, he notes, with astonishing regularity, the ailments for which he and his wife, Josephine, sought treatment from local doctors—dizziness; strange inflammations; "bilious attacks"; headaches; chin pain; agonized feet, hands, elbows; catarrh; rheumatism; lumbago. Illness was a matter of course, survival often a matter of luck.

If the diagnoses were suspect, the treatments were even worse. When cholera was detected in the early days of the Santa Fe Trail, doctors first advised bed rest, and if that didn't work, they administered calomel—mercurous chloride—a powerful "cathartic" that caused diarrhea, vomiting, excessive salivation, and hair and tooth loss (as if the violent catharsis brought on by the cholera bacterium weren't enough). If that failed, they'd soak victims' hands and feet in scalding water containing mustard and common salt. Next came bloodletting to "clear out" the infected blood—using (not at all sanitized) syringes, lancets, or leeches applied to veins in the forearms, neck, or temples, or a terrifying spring-loaded device called a "scarificator." They then dispensed a few spoonfuls of oil of vitriol, a concentrated and highly poisonous sulfuric acid compound. Then came an enema of chicken tea with a tablespoon of salt. If the patient was still alive, they would next try a Hail Mary concoction of calomel plus camphor plus quinine plus morphine. Next, typically, came coma, and death.

The historian Merrill J. Mattes estimated that one of every twelve immigrants died on the overland trail; the presence of doctors did little to improve the odds. In fact they probably made them worse. Nor had treatments improved all that much in the two and a half decades since Julia had first arrived in Santa Fe.

No wonder she sought treatment in Germany.

◇◇

After an initial consultation in Berlin with Olshausen, Julia returned to Hanover. There, she visited with her mother and four of her sisters—Adelheid, Sofie, Regine, and Auguste. A cousin, Paul Schuster, called on the Staab women in their hotel almost every day. "We like him very much," Bertha wrote, "he is lively and quite good looking." Bertha noted, too, that Tante Auguste gave her a spoon. Souvenir-spoon collecting, I learned, was a fad in the late nineteenth century; they were collected like postcards, or Beanie Babies.

On Julia's next visit to Dr. Olshausen, Bertha accompanied her, diary in tow. "Mamma has had an operation," Bertha wrote, "performed in Berlin." Bertha made no mention of what the operation was or what condition it set out to treat—perhaps such things were simply too delicate to discuss in that era, even in the private pages of a journal. I suspect, however, that it was a hysterectomy, since that was the surgery for which Olshausen was most famous.

It is clear that Julia was unwell, and that her problem was gynecological. But this still tells us little about her suffering, because in the late nineteenth century, nearly everything involving female health was considered gynecological. All sorts of ailments—and especially emotional ones—were lumped in the general category of "female problems." The word "hysteria"—that diffuse mental diagnosis common among genteel white women in the nineteenth century and characterized by convulsive fits, trances, and tearing hair—derives from the Greek word "hystera," which means uterus. The womb was linked indelibly in the medical mind to mental illness. Plato had believed that a woman's uterus roamed the passages of the body like a ghost, unleashing emotional disturbances.

Nineteenth-century diagnoses were more clinical than Plato's wandering womb, but they were essentially similar: the Irish physician Thomas More Madden, in his 1893 book *Clinical Gynaecology*, described female mental illnesses as "the reflex effects of utero-ovarian irritation," while Joseph R. Buchanan, a mid-nineteenth-century American neurologist, argued that the woman's womb was her "region of insanity." "The organ of baseness," he wrote, "lies along the posterior margin of the abdomen, and between the ribs and ilium, connecting above with irritability and below with melancholy, through which it approximates the region of mental derangement." An illustration from his book, *Outlines of Lectures on the Neurological System of Anthropology*, superimposes the map of that terrifying region on a lithograph of Praxiteles's *Aphrodite of Knidos*. Dr. George Beard, the nineteenth-century American neurologist who invented the diagnosis of neurasthenia, suggested that all women with mental problems should undergo a gynecological exam.

Thus, hysterectomies such as the one that Dr. Olshausen may have performed on Julia were expected to alleviate not only diseased wombs but also distressed female psyches. Doctors removed the womb to treat hysteria, "excessive female desire," as well as neurasthenia, another nervous disorder common to middle- and upper-class women in the nineteenth century. The symptoms of that disorder were even more diffuse: they included languor, sleeplessness, nightmares, headache, noises in the ear, heaviness of "loin and limb," palpitation, flushing, fidgeting, "flying neuralgia," spinal and uterine irritation, hopelessness, claustrophobia, germophobia, and general morbid fear.

It is hard to know, then, if Julia's surgery was aimed at her physical or mental ailments, since the suggested treatments for both were essentially the same: gynecological treatment or what was known as "the rest cure"—in Julia's case, both surgery and the rest cure were prescribed. The latter involved bed rest, seclusion, a bland diet, and the renunciation of all intellectual pursuits. Writing, painting, drawing, and education were considered too stimulating for the mentally infirm female, as was reading novels, especially exciting ones. Women were counseled instead to read books on practical subjects, like beekeeping. The idea was both to calm the nerves and to regress the patient to a receptive, infantile state. Doctors also prescribed the rest cure for tuberculosis, arthritis, asthma, and bad hearts, but they were particularly fond of its use in the treatment of mental illness.

The nineteenth-century author Charlotte Perkins Gilman took the rest cure for depression after the birth of her first child; her doctor forbade her from writing, sketching, or reading, and, she explained later, she was reduced to playing with a rag doll on the floor. "I went home and obeyed those directions for some three months," she wrote in a 1913 magazine article, "and came so near the borderline of utter mental ruin that I could see over."

At the extreme end of the rest cure, institutionalization was an

option; Julia was fortunate to avoid this. In 1887, Nellie Bly, the cru-
sading newspaperwoman, feigned madness to expose the torturous
treatments imposed at the Women's Lunatic Asylum on Blackwell's
Island in New York, a hospital that served mostly immigrant women.
Quickly diagnosed—"Her delusions, her apathetic condition, the
muscular twitching of her hands and arms, and her loss of memory all
indicate hysteria," explained the admitting physician—she was made
to sit in the cold for much of each day, fed spoiled beef and tainted water, and doused daily with buckets of cold bathwater. "Take a perfectly
sane and healthy woman," Bly wrote in her subsequent exposé, "shut
her up and make her sit from 6 am to 8 pm on straight-back benches,
not allow her to talk or move during these hours, give her no reading
and let her know nothing of the world or its doings, give her bad food
and harsh treatment, and see how long it will take to make her insane."
Her newspaper, *The World*, arranged her release after ten days. Other
inmates weren't so fortunate. For that reason the wealthy, like Julia,
preferred to undergo treatment in their homes.

Julia was, in some respects, lucky. She had money, and a family—
including, yes, a husband—who worried for her and supported her,
though Abraham, like most men of the time, did so mostly from afar.
Depression is lonely and often terrifying now, but back then there was
even less recourse for those in its grip. There were no antidepressants,
no cognitive-behavioral treatments; there was no talk therapy or Va-
lium. Julia could have squeezed into Dr. Scott's electric corset, a girdle
fitted with steel electrodes where the whalebones should have been, or
tried other forms of "electrotherapy" intended to recharge a patient's
enervated nervous system through the application of low-voltage current.

She could have taken patent medicines—dubious, sometimes dangerous herbal concoctions. Dr. R. C. Flowers's Nerve Pills were designed to "Overcome Sleeplessness, Restlessness, and Hysteria," while

"Dr. Pierce's Favorite Prescription" was supposed to build up "the shattered nerves" by acting "directly on the delicate and important organs concerned in wifehood and motherhood." An 1890s advertisement in a Santa Fe newspaper for Dr. Pierce's remedy urged readers to imagine full recovery. "The fan that long lay listless and idle in the lap of an invalid again speaks the eloquent language of a healthy, happy woman." Julia probably did not have access to a treatment available in East Coast cities, in which doctors massaged women's genitals to elicit therapeutic "convulsions." This would have been a pleasant course of treatment, no doubt. So, too, would visits to a spa for a dose of healing and rest.

And this is what Julia did after a week's recovery from her surgery in Berlin. It was time for gentler treatment. So she and Bertha boarded a train heading west and slightly south, past Hanover, past Hameln, to an elegant spa town called Bad Pyrmont.

sixteen

❧ LOW SEASON ❧

The Montezuma, New Mexico, hotel and hot springs, 1888.

A hundred and twenty-odd years later, I followed in Julia's and Bertha's footsteps—this time seeking not a cure , but a ghost. I boarded a plane, not a train, to Bad Pyrmont, and it wasn't July but late October—low season. Like Bertha, however, I brought my mother—not because she suffered from any noticeable neurasthenic ailment, but because she speaks German.

My mother was a good sport in abetting my quest for Julia. She thought my ghost-hunting efforts amusing, if silly. It was the search for my father's German family history that lay closer to her heart. She had studied German for many years as a young woman. I had studied it for a few weeks when I was in kindergarten and had learned only to count to six. So I deputized her as my translator and travel companion and dragged her along on my hunt. Our first task was to meet a local historian named Manfred Willeke.

Herr Willeke—we never addressed each other by first name— lived in Lügde, Julia's Lügde, which was only a few kilometers from Bad Pyrmont. By happy coincidence, he was the designated historian for both cities, and he agreed to help me trace Julia's path in both. Arranging to meet with Herr Willeke had had its roadblocks, however: he spoke little English, and I spoke even less German. After making contact, I wrote an email to him in English confirming our appointment. He responded in German, asking why I insisted on visiting on a day when he was busy. "Why have you not taken this into account?" he wrote. "It is a pity that apparently everything is so strange and does not seem to fit when I wanted so much to help you. . . . I doubt that we'll see you at all."

"No, no!" I wrote back, in German this time, with the help of my mother. "I'll come whenever you can see me!" Our relations from then on were quite cordial, provided they were in German. He signed his emails thus: *Herzliche Grüße aus dem Tal der sprudelnden Quellen*— Warm greetings from the valley of bubbling springs. The valley of Bad Pyrmont—Lügde's valley, Julia's valley: what a lovely spot it portended.

Germany had never before appealed to me as a vacation destination. I always pictured a gray and industrial landscape, flattened by war and brutalized by modernity. My mother's father, a Baltimore Jew of eastern European ancestry who had watched the Holocaust from afar in horror, disapproved of all things German. My grandfather refused to buy German cars, or chocolate, or anything else from that hated place, and he tried, rather pointedly, to steer my mother from studying German in high school—he urged her to take French.

But she loved the solidity of the German language—the clear and structured grammar, the way the words were contained in little consonantal boxes. And it was also a not-too-dramatic way to prove herself a rebel. So she went ahead and studied German in high school, and then in college, and when she graduated she moved to Germany. Back home in the States, my mother studied yet more German in graduate school, reading Lessing, Goethe, Rilke, and Brecht in the music of their own language. My own act of rebellion, perhaps—a very small one of many I inflicted on my mother—was to have no interest in anything remotely German *or* Jewish. Until this ghost hunt.

I left home with some trepidation. I was excited to learn about my family, but also nervous that I would find no information at all about Julia and Abraham and the world in which they had grown up. It would also be the longest time I had been away from my children, then five and two years old. Of course the journey was nothing like what Julia had undertaken when her children were similar ages—no stagecoaches or trains or weeks-long steamer journeys across the ocean, leaving some of her children with nannies, taking others with her.

But Herr Willeke was right. There are lovely spots in Germany—many of them. The countryside there has a restrained and tended beauty—the towns and cities, too, with their meticulous flower boxes and carefully considered architecture. Bad Pyrmont, in 2012, wasn't all that different in appearance from the high-baroque city that Julia had visited more than a century earlier. It was a lovely, well-ordered town, full of old colonnaded buildings and cobbled roads. Square neoclassical hotels lined the streets, along with ornately gabled boardinghouses, all quite grand, if slightly gone to seed. Elderly German "healthwalkers"—angular and white-haired, in brimmed hats and beige parkas—promenaded carefully along the wide boulevards, past the doctors' offices cum gymnasiums, and through the enormous baroque *Kurpark* at the edge of the city center, the gardens all soothing straight lines and symmetry: flower beds, ponds, arched bridges, weeping trees, palm gardens, topiary, a moated castle. "Mental health is body health," we read on a placard.

It wasn't hard to picture Julia and Bertha wandering the same wide, well-tended paths, seeking health and consolation in the manicured woods of the *Kurpark*—another mother and daughter trying to reconnect, each in her own way, to a diffident *Mutterland*. On our first night, my mother and I wandered the restaurant district browsing menus. It was a homecoming, of sorts, for both of us—my mother trying to recapture her fluency in German and relive her student years there, me trying to understand the world my ancestors had inhabited so many years before. We were having fun, my mother and I, and it made me feel even more poignantly how the pages turned toward the end of Bertha's diary, the time ticking down toward Julia's decline and death five years hence.

Though Bad Pyrmont had once attracted tourists from all over the world, it mainly hosted elderly Germans now, big-boned and red-cheeked. We stood out: my olive-skinned and petite Jewish mother, with a dramatic streak of gray-white hair framing her face and matching gray glasses, and me, slightly fairer, taller, an American mishmash of culture and blood. We wandered past German restaurants, generally empty, a few Italian places, generally packed, and a Greek restaurant, where we ate food that resembled a German fantasy of Mediterranean cuisine: feta cheese, but also cabbage and the ubiquitous *Schwein* in various configurations. Julia and Bertha wouldn't have known what to do with such crossbred cuisine.

After dinner, we wandered into a *Nachtclub* where a band was playing, though I thought at first it was a karaoke act. A doughy German woman with black-dyed hair belted out a disco version of "Volare" to the graying crowd, and she harmonized not altogether well with a Turkish fellow in loose soccer clothes. Phantoms of smoke furled at eye level. The city was so well preserved, in both its architecture and its intent, that I half expected Bertha and Julia to wander in and perch full-skirted at the next table, expressing their displeasure with the barbaric harmonies coming from the stage.

Julia had gone to Bad Pyrmont to heal; I'd come there to discover what inside Julia was broken. While I wasn't sure what I could learn, the city, with its old cobbles and healthwalkers and pedimented buildings, made it easy to dwell in the past. With Bertha's diary in hand, I half lived there already. Our hotel, the Fürstenhof, helped in this regard. A yellow, block-long early baroque establishment, the Fürstenhof had stood just off the city's main square since 1777 and felt little changed: threadbare carpet, peeling paint, rickety elevators, surly front-desk employees. It was, like everything in Bad Pyrmont, elegant if spartan and a bit bedraggled. But there was a familiarity to it—from knowing that it had been there when Julia visited, and that my great- and great-great-grandmothers may have dined in the same hotel restaurant where we planned to eat breakfast with Herr Willeke the next morning.

We did eat breakfast with Herr Willeke the next morning, though only after further miscommunication—we waited to meet in the lobby, he in the breakfast room. Herr Willeke was in his mid-forties, but his hair was already white, thick, and arranged in a neat pompadour. He wore a navy blazer with a pocket square and an open-collared shirt, and there was an antiquarian sense of displacement about him—as if he, too, inhabited the past. He ushered us into a 1959 Audi that had been his father's, and drove us up above the springs to the town archives. Herr Willeke spoke briskly, turning around frequently as he drove to make sure I was listening. My mother translated furiously from the front seat while I sat white-knuckled in the back, willing Herr Willeke's eyes back on the road. He wove the Audi through the ever-narrower streets that climbed from the city center and stopped, finally, at a neo-Gothic spire: the town library.

Herr Willeke took out a large set of keys and ushered us in. The archives resided at the tippy-top of the tower, and we curved up an elaborately vaulted spiral staircase plastered with German Harry Potter posters until we arrived at a sunlit room that smelled of old papers. Herr Willeke extracted from the shelves a large bound book containing many years' worth of the city's *Kurliste*—the daily "cure list" for the spa.

I knew that Julia and Bertha had traveled to Bad Pyrmont in July. "Left for Pyrmont July 1," Bertha had written in her diary. So we pulled out the book that covered the summer of 1891 and began leafing through brittle pages that listed each visitor's name, date of arrival, place of origin, and place of lodging. The guests mostly came from Germany—Berlin, Bremen, Hanover, Munich, Düsseldorf—but also from Chicago, New York, and St. Louis. In the July 15 edition, across from advertisements for Suchard Chocolat and Leibig's Fleisch-Extract—drinkable beef syrup imported from Uruguay—we found *Frau und Fräulein Staab*, hailing from *Neu-Mexico*. Julia had arrived in Bad Pyrmont, and was there, lurking in those pages, waiting to be healed. And I realized that I was hoping, somehow, that she could be.

The specific treatment offered in Bad Pyrmont involved hydropathy, known more commonly as the "water cure." Though baths and mineral springs had seen therapeutic use dating back to ancient Egypt, this particular cure had grown fashionable among the upper classes and their doctors in the eighteenth and nineteenth centuries. If time and money allowed, patients "took the waters" at the famous spas of Europe—Carlsbad and Wiesbaden and Baden-Baden, Marienbad and Gastein, and also Bad Pyrmont. In the American West, wealthy health-seekers favored the Hotel del Coronado in San Diego, where Abraham and Bertha had stayed during their California sojourn; the Atchinson, Topeka and Santa Fe Railway had also built an elegant "hotel for invalids" in Montezuma, New Mexico, at natural hot springs about seventy miles from Santa Fe. Julia and Abraham visited those springs regularly. "They are a sure cure for chronic diseases," wrote the Denver *Rocky Mountain News*, "such as rheumatism, gout, scrofula, and other diseases of the skin, especially syphilis."

The idea was, essentially, to wash away disease: "Wash and Be Healed" was the spa movement's motto. Patients would expunge bad air, impure food and drink, indolence, overexertion, improper light, and unregulated passions through quaffing and bathing, and then quaffing and bathing some more: sitz baths, eye baths, hand and foot baths, pouring baths, half baths, full baths, hot baths, warm baths, cold baths, head-dousings, hot compresses, warm compresses, cold compresses, mineral water consumption, mineral steam inhalation, and cold water immersion of the unpleasant sort that Nellie Bly and the other hystericals received at the lunatic asylum. ("Place the head over a basin, and pour water from a jug over the head and chest until the patient becomes chilly and revives," wrote the hydropathic pioneer R. T. Trail.)

The water cure wasn't only for the weak; the hale and hearty also thought it beneficial. The feminist icon Susan B. Anthony took the waters in Massachusetts in 1885. "First thing in the morning," she wrote, "dripping sheet; pack at 10 o'clock for forty-five minutes, come out of that and take a shower, followed by a sitz bath, with a pail of water at 75 degrees poured over the shoulders, after which dry sheet, and then brisk exercises. At 4 p.m., the programme repeated, and then again at 9 p.m." Each day at the spa involved "four baths, four dressings and undressings, four exercisings, one drive and three eatings." In Bad Pyrmont, it involved many drinkings as well: Herr Willeke told me that visitors drank sixty or so glasses of water in two hours.

Different minerals were supposed to possess different healing qualities. Alkaline waters were recommended for diabetes, malaria, and reproductive and genitourinary afflictions; salty waters were best for skin afflictions and catarrhal (phlegmy) infections; sulfur was good for liver and respiratory disease. Five thousand meters below the placid city where my mother and I sat with Herr Willeke, poring over old advertisements for chocolate and patent medicines, more than seventy water sources mixed with those minerals and erupted from a volcanic fault. In the cast-iron temple in the center of town called the Wandelhalle, one could drink from an array of the town's most cherished sources. The Helenquelle, the Helena spring, emitted a *Sauerwasser*—salty water—that was good for digestion, but bad for the heart. The Friedrichsquelle contained iron and was good for the heart. The Trampelquelle helped with digestive distress; the Augenquelle reduced eye inflammation. Water from those springs was also piped into the steel bathing house and diverted into individual bath chambers and mist machines.

The Bad Pyrmont waters were used to treat anemia, "weak blood," malaria, neurological problems, impotence, bedwetting, "albumin in the urine," a "calloused heart," "English sickness," and "masculine sexual malfunction" such as excessive sperm production and "shrinkages." They were thought to be particularly helpful for problems like Julia's—"the so-called women's conditions," wrote a Dr. Seebohm, who published a travel guide touting Bad Pyrmont around the time Julia visited: "blood and mucous flows, irregular and painful periods, problems during pregnancy and post-partum, infertility." That July, Julia spent weeks in bathtubs, sopping up the waters. What she suffered, mental or physical, the doctors of Bad Pyrmont believed they could relieve.

◇◇

Along with all the dousing and soaking, Bad Pyrmont offered plenty of socializing. Bertha was excited at the prospect. German spas, frequented by royalty and the growing class of people who lived like royalty, played an important role in the elaborate social rituals of the wealthy. Princes and hangers-on passed through Bad Pyrmont regularly. Peter the Great visited in the eighteenth century; Crown Prince Friedrich Wilhelm of Prussia was said to have accidentally invented the trouser crease while on holiday there in the nineteenth. The iconic German poet Johann Wolfgang von Goethe visited as well, finding Bad Pyrmont's vapor caves mesmerizing: "the soap bubbles happily dancing on the invisible elements, the sudden extinguishing of a lambent straw smudge, the instant ignition of a candle." Tens of thousands of people came there each summer.

Indeed, Herr Willeke thought it impossible that Julia and Bertha could have stayed in Bad Pyrmont for longer than a week. It was simply too crowded to accommodate lengthy stays, he told us, and too expensive. I knew from the diary, however, that Julia and Bertha did stay for longer. They had the money, clearly, and also the motivation: they were desperate to get Julia better. Intrigued, Herr Willeke looked again at Julia's entry on the *Kurliste*, running a stout finger down the musty column that announced the guests' sleeping arrangements. When he found Julia and Bertha's lodgings, he clucked in excitement. They had taken rooms at the private pension of Frau Alice Breithaupt. He knew exactly where.

We descended from the library tower, clambered into the Audi, and threaded the streets to a hillside overlooking a small park. The pension of Frau Breithaupt—Julia's home during her stay in Bad Pyrmont— was an oddly shaped three-story building at the intersection of two angled streets. Ivy climbed the gray plaster facade, which was punctuated by an arched door, elaborately linteled windows, and a curly wrought-iron balcony. A century and a quarter after Julia's stay, a restaurant now occupied the ground floor, and the offices of a financial adviser and a mediator resided on the second floor. The top floor still housed dwellings, their large windows full of light. I imagined Julia standing at the third-story balcony, the valley spreading out beneath her, breathing the healing air. It was a good location, Herr Willeke assured us—close to the *Kurpark* and the baths. We must have looked a strange group, staring up at this old building, Herr Willeke in his timeworn blazer, gesturing hurriedly, me with my notebook and smartphone, taking notes and photos, my petite gray-haired mother bouncing between us, translating gamely. We walked around the side of the building for another view, and Herr Willeke pointed at the building just downhill from Frau Breithaupt's. He told us it served during Julia's time as a home for illegitimate children, produced by the couplings of spa guests and the local servants.

There was plenty of coupling in Bad Pyrmont, he said, illicit and otherwise. These days, the odds were much better for those in Herr Willeke's set. "It's a good place to find wealthy widows," he joked. Back then, however, the spa was a marriage market for the young: a place not only to heal, but to mingle. "It was very common," Herr Willeke told us, "that girls came there looking for husbands." Bertha, alas, wasn't able to mingle nearly as much as she would have liked. She was stuck alone with her invalid mother. "Mamma is not able to go out much—Do not know a soul," she wrote. Thus she "was awfully glad" when Abraham and Delia came from Carlsbad to join them a week after she and Julia arrived. Other family—Bertha's spinster aunt Regine and her aunt Bernhardine and uncle Bernhard Nussbaum—also visited from Hanover. Julia appeared cheered by her siblings, Bertha wrote; they were part of the cure. Some of them, however, irked Bertha. "I cannot bear Uncle N"; Uncle Nussbaum, Bertha wrote, "grunts all the time and thinks everything I say funny—not silly but foolish." She could bear his son Arthur better. "Took walks with Arthur N to the Erdfalle and Bomberg—very pretty walks."

Besides the relentless promenading along the Kurstrasse—bonnets and bustles and ruffles and gloves, full regalia—there were, during high season, garden parties, balloon rides, fireworks, three Princely Cure Ensemble concerts each day, and five theater performances each week. Bertha attended a number of shows. "Am gone on an actor who plays in the Pyrmonter Theatre—Very good-looking—Hermann Leffler," she wrote—he would later become a star of Berlin theater and early German film. "He hasn't vouchsafed me a glance, however." Even Bertha's foolish starstruck moments seemed momentous to me, reading them a hundred and thirty years after they were written; they were antediluvian in their language but somehow modern in their sentiments.

Bertha also managed to attend some of the Saturday promenade balls in the Kurhaus reception hall. "Fri 24th July—mamma was sick," she wrote in her diary, alarmed that she might not be able to attend that week's ball. Julia's progress in recovering from her surgery had proceeded slowly. By Saturday, however, she seemed well enough that Bertha and Delia could attend the ball. But it was a disappointment: "No gentlemen came near us. The proportion of ladies to gentlemen was ten to one." So the Staab girls, brashly, made do. "Delia and I danced together, also with a Fräulein Kasse from Berlin—and thereby astonished" the dignified Europeans who were present. In America, their propriety seemed dowdily German; in Germany, they were daring and American.

seventeen

❧ SCHUSTERGARTEN ❧

In Lügde's Jewish cemetery.

Having exhausted Bad Pyrmont's Julia-related archival resources, we climbed back into Herr Willeke's ancient Audi and set out for Lügde. The road from Bad Pyrmont to Lügde barreled straight alongside the autobahn, past car dealerships, factories, and warehouses, until the highway tunneled under the old town of Lügde. From there we left the larger road and curved up a hill, winding up to an ancient Romanesque stone church built on the spot where Charlemagne celebrated Christmas in the year 784. Herr Willeke parked the Audi on a gritty shoulder and we strolled over to the church. It was cold inside the thick stone walls, the late autumn light chinking in through a few high windows. Outside, Herr Willeke told us about a rosebush said to bloom eternally over an image of the Virgin Mary found in a stone on the site; now an equally miraculous eight-hundred-year-old linden spread its limbs above us. We would see more of Lügde, Herr Willeke promised, but first: lunch.

He drove us to a hotel in the hills above town, worth it for the view alone—a patchwork of green, the river Emmer lazing past, narrow roads that wound through tunnels of beeches and lindens, past apple orchards, stone farmhouses, handsome feudal villages, and picture-book forest, weeping trees in autumn opulence. The hills ascended gently toward Köterberg—Mount Bow-Wow. How Julia must have loved the place this time of year. As I gazed out across the lush scenery of Julia's childhood, I found it easier to understand what she must have felt was taken from her.

In the restaurant, Herr Willeke ordered carefully. He had thin, white wrists, and delicate hands that seemed not to fit his otherwise meaty frame. He was, he told us, allergic to all sorts of things—nuts, honey, milk, beer—everything except meat and processed foods. He sang in a choir and he had run, several times, for local office as a Social Democrat—but it seemed to me that the past was, somehow, more real to this beer-sensitive German than the present moment in which we sat looking over the Pyrmont valley and Lügde, eating *Schwein*. There was a wistful formality about Herr Willeke. We wandered the same bygone world, he and I, sifting through ghosts, mining the past for clues to the present.

On the way down from lunch, we wound through the new section of Lügde, built after World War II to house German refugees from what was originally Silesia, now part of Poland. Other migrants, mostly

Turks, lived there now, their ersatz *Fachwerk* half-timber homes cling-
ing to the hillsides. Herr Willeke pointed out seven small chapels—
stations of the cross—and the place where the flaming *Osterrad* hay
wheel still rolled downhill each Easter, as it had for hundreds of years.
"It was a tradition which the Schusters"—Julia's family—"would have
known," he said. I imagined the Schusters gathered to watch: Julia as a
child with her many siblings; Julia bringing her own children from the
New Mexico desert to see the wheel retracing the same path year after
year, a tradition that would outdate and outlive them all. It was odd, to
feel nostalgic for a place I'd never been and a tradition I'd never witnessed, but I did. There are places that always bring a flood of childhood ghosts: the stunted forsythias at the entrance to my elementary
school; the crackling ponderosa forest behind my great-grandfather's
stone home in the New Mexico foothills. On this redolent autumn day,
Lügde swamped me with a similar wistfulness.

We climbed back into the Audi and descended toward the old village, parked at the edge of what was once the upper city gate, and wandered uphill to a fenced-off park sandwiched between a housing project
and an elementary school. Herr Willeke led us to a wrought-iron gate
hung between two ivy-mounded brick posts and adorned with a Jewish star. This was the Jewish graveyard—the Schustergarten, they
called it, because the town's Jews had purchased it in 1887 from Julia's
mother, the widow Schuster, in order to relocate an earlier plot beside
the town wall. We strolled across well-tended grass toward the stone
grave markers: some upright, some listing. The first stone we reached
was that of Philipp Schuster, who died in a 1866 cholera outbreak that
killed "126 people and one Jew," according to town record-keepers.
Philipp was that one Jew—Julia's cousin, who died the year that she
arrived in America, his stone a reminder that in 1866, in Lügde, a Jew
was not considered entirely a person. Julia's family—and Abraham's—
came from here, but they never belonged.

We found Julia's father's gravestone as well: Levi David Schuster, who died on May 16, 1877. That was, I realized, only two weeks before Sister Blandina accompanied a troubled Julia by horse-drawn coach to Trinidad so she could take the train east to New York and travel from there to Germany. Julia's father had only just died. The trials of motherhood and "female problems" and loneliness may have helped form her sadness in the days when Blandina cared for her, but something much more specific also played a part: an ocean and a continent away, Julia's father was gone.

Even after all the months I had spent digging through her past, Julia remained so tantalizingly remote. I was coming to know the people around her—Bertha through her diary, Abraham through his public presence. But Julia, sequestered by time and disposition, continued to elude my understanding. Now, however, I had learned that she had lost a parent. This sadness I could understand. Of course she was unhappy; she was mourning.

We passed some graves so heavily mounded in ivy that it was impossible to find the stone—Herr Willeke, who kept a map of the graveyard in his house, assured us they weren't Schusters or Staabs. In the back corner, we found a familiar name, "Jette Staab, geb Spiegelberg"—Abraham's mother, who had married Moses Staab in 1832. She was a Spiegelberg by birth, which explained, a little better, how Abraham came to Santa Fe. He and the Spiegelberg brothers must have been first cousins; his mother was a sister to their father. The stones told us this much.

We climbed back in the Audi and drove slowly now, inside the city walls, along a cobbled street to Herr Willeke's home, a classic frame-and-timber house that had been occupied by his family for more than two centuries. There was a gingerbread humility about Lügde's houses, with their *Volk* writing on the beams above the entry doors— Herr Willeke's offered the information that it had been built in 1806 and renovated in the 1860s; a door lintel down the road took a more confrontational approach: "May all my enemies die suddenly and come back as ghosts." Lügde was full of ghosts, Herr Willeke told us with a quick smile. And I could feel them: if not the specific ghosts, then at least the weight of the past.

We wandered two blocks down Herr Willeke's street on foot, into the center of town. A bridge ran across the Emmer, its banks stacked with riprap against the summer floods. A park and a playground occupied the upriver meadows; we could see an electrical wire factory a few hundred yards downriver. We headed back up another cobbled street of large *Fachwerk* homes—all thirty meters long, many built in the 1600s. Their plaster-and-wood patchwork facades were reassuring in their Old World sameness, house after house with steep shingled roofs and big arched front doors.

We walked around Lügde's cathedral—it was tall and dignified, of brown brick, a bit dark. It had been built in 1895, Herr Willeke said, and during the construction, the town's Jews had given money for the rose window in the front. "This was not unusual across Germany," he told us; Jews often contributed to the construction of Catholic churches in the villages and cities where they hoped to remain in good graces. The cathedral had replaced an older church that stood during Abraham's and Julia's childhoods. And above the door of that church, Herr Willeke said, four letters were chiseled in Hebrew: JHWH, Yahweh.

Abraham had been familiar with tetragrammaton engravings on churches. He had known, well before Archbishop Lamy ever built his cathedral in Santa Fe, that they were not uncommon. Abraham was not "totally ignorant" of the fact that those Hebrew letters adorned Catholic churches across Europe, as Rabbi Fierman had posited. So perhaps he had funded the cathedral after all; perhaps he had asked Lamy to place the Hebrew letters above the door as a symbol of the contribution of Santa Fe's Jews to the city's spiritual life and material foundations, and also, perhaps, as a reminder of the church of his and Julia's childhood.

We walked past an ice cream shop that occupied a house where some Schusters had lived long ago, and watched two small Turkish children playing tag. Around the corner, we found the home where Julia had lived as a child. It was another *Fachwerk* house—large, with brown timber and white plaster, many symmetrical windows, a tall peaked roof, and an arched door. An air-conditioning unit stuck out incongruously on the right side—not every house stands unaltered as a monument to the men who built it. It was now a clothing store, Herr Willeke explained, hands aflutter. The store was closed, so we couldn't go in—it wouldn't have resembled the home of Julia's childhood anyway.

Back at Herr Willeke's house, in the upstairs sitting room that served as his archive, we pored over documents he had copied from the *Stadtarchiv*. We had climbed a steep set of low-ceilinged stairs and wound through many small rooms to get to the sitting room, where every available patch of wall space, every shelf, was covered in bric-a-brac: putti, knickknacks, portraits of family members in baroque frames. Herr Willeke served us tea and showed us his files. The Jews of Lügde had, of course, been erased as thoroughly there as they had everywhere in Germany, but their presence remained in the village that once abided them, and also in Herr Willeke's papers—as letters and numbers, words and names. We spent the rest of the afternoon at his coffee table examining stacks of paper that Herr Willeke had diligently, tenderly copied from the archives: census records, a map of the graveyard, chronicles of Lügde's Jewish families, whose histories ended mostly in America if they were lucky, or in the camps and ghettos of eastern Europe if they were not.

The light was fading as Herr Willeke drove us home. We had spent a long day of searching for people long dead. But with the help of Herr

Willeke—so formal and prickly, and also endearing—I had found something of Julia in Lügde. They were small clues, quiet whispers from the past. Her home did not stand apart like the mansion in Santa Fe; it resembled the homes to the left and right, up and down the street. The chronicle of the cholera epidemic in the town records reminded me starkly that the year Julia left for America, a Jew was still less than a person in her hometown. And the moss-and-ivy-draped stones of the Schustergarten informed me that, far from home, Julia had learned of her father's death and set off, through wild lands rampaged by Billy the Kid, to pay her respects. We had unearthed in that Lügde cemetery one small fragment of her unhappiness.

What we had learned of Abraham was less affecting, and it wouldn't solve the debate as to how the Hebrew letters came to be carved above the Santa Fe cathedral door. But Herr Willeke's description of the old Lügde church had given us this much: Abraham could hardly have been ignorant of the custom of the tetragrammaton. And perhaps he, like his wife, also valued some reminder of his village past.

The next morning, the good Herr met us in the Fürstenhof breakfast room and drove us to the train station. His propriety in squiring us through Julia's world was touching; he might have been a Bad Pyrmont gentleman of Bertha's day. As we detoured past the Schloss Pyrmont, the large stone-and-plaster castle that loomed above the *Kurpark*, he repeated a saying: "Those who drive past the castle will always come back." He regarded me for a long moment in the rearview mirror. "But don't come back as a ghost," he said. "We have enough ghosts."

eighteen

❧ THE MERCHANT PRINCE ❧

Abraham in his prime.

After nearly a month spent tending to her mother, Bertha traveled to Switzerland with Abraham, leaving Delia, back from her own jaunt with their father, in Bad Pyrmont to watch Julia. Now it was Bertha's turn to explore. After five train changes and one Russian lamb cape left behind in a railroad car, she and Abraham arrived in Lucerne, "hungry as prairie wolves." The next day, she saw the Alps for the first time: "It was a grand sight—the bluish-greenish water and the snow-capped mountains looming above it." Bertha liked the Swiss people she met; she found them "courteous." Freed from her mother's infirmity, she behaved like a tourist rather than a health-seeker: sightseeing, visiting springs and lakes and castles and spas, hopping from hotel to hotel, shopping for dresses and spoons, and hiking among the sublime peaks on the Swiss-French-Italian border. ("Went for steep hike down slippery descent, Papa said, 'It's a wonder you didn't fall.' I kept my peace and didn't tell him that I had fallen—his back was turned however and I picked myself up as fast as I could and walked on with an unconcerned face.")

Abraham continued to take "baths and douches" for his health, and he continued to spoil his youngest daughter; Bertha persisted in striking up a conversation with any "young gentleman" who crossed her path. She expressed dismay at the local table manners ("Shaking hands over the table and reaching for dishes in front of the other guests are common practice—Finger bowls are not to be had.") She beheld the *Alpenglühen*—the setting sun reflecting on the peaks, the mountains "all afire."

In mid-August, Bertha and Abraham left for Frankfurt, where they attended an electric exposition—"mostly machines, very interesting for those who understand the mechanism." For Bertha, the motors, lightbulbs, and electric waterfall, powered by a hydroelectric station one hundred miles away, were as mysterious as the "hidden power" she'd observed in Annie Abbott. Bertha also made a visit to Goethe's house, where she accidentally spurned a Balkan prince. "The prince of Montenegro was there at the same time I was and asked me in French whether I could speak that language," she wrote. "Having had a lecture from Papa in the morning not to be familiar with strangers, I answered him"—the prince—"in as few words as possible and went my way. Afterwards on walking to the register saw written down Prince de Montenegro Nicolai and was told that the gentleman who had spoken to me was the one. Tra-la?"

On Tuesday, August 18, they left Frankfurt for Bad Pyrmont, and Bertha's frolic ended. Julia and Delia met them at the depot. "Mamma is not well," Bertha wrote—as she had written before, as she would write more and more as the weeks passed. The following day, Abraham boarded a train for Hanover. "Papa went . . . to see about rooms for us—see if mamma feels better there." They joined him, taking rooms at the Hotel Royal, but Julia did not feel better there. "Mamma cannot stand the noise," Bertha wrote. Abraham and Delia went to scout a health resort in the Harz Mountains of northern Germany. The Harzburg resort was a sleepier spa with saltwater thermal springs. They hoped Julia might find enough peace and solitude there to heal.

Julia felt better on the day they packed their bags to depart from Hanover. She was looking forward to a change of scene, away from the clamor of the city. The family visited Julia's mother, the widow Schuster. She gave Delia a beautiful spoon and Bertha "a lovely chatelaine"—a waistband clasp. A slew of aunts and cousins accompanied their New Mexico relatives to the depot. "The farewell was exciting," Bertha wrote, "we all cried." Not so much, though, that Bertha was unable to keep an eye out for suitable fellows. "I saw a young gentleman look at us as if we were curios and smile—we must have looked funny."

The farewells were also for Abraham, who would not accompany them on the next leg of Julia's therapeutic journey. "Papa leaves Bremen for New York on the 25th."

He left Julia in the hands of his daughters and headed back to America, to New York and then Santa Fe—to the world of commerce and men, where he knew what to do and how and when to do it. These things were much easier than trying to repair his ailing wife.

◇◇

Even as he struggled to stanch Julia's decline, Abraham remained vigorous. After Zadoc's death in 1884 in Carlsbad of the "liver affection," the Staab company had grown only stronger, the family only richer. The frontier suited Abraham's sensibility. He would not have agreed with the assessment of another Jewish immigrant, Phoebus Freudenthal, who had written from Las Cruces, in southern New Mexico, to tell his brother in Germany that "the only thing New Mexico is good for is making money." Freudenthal figured that the rugged country of the Southwest must have been one of the last places that God created: slapped together thoughtlessly and hurriedly. It's possible that Zadoc and the Spiegelbergs shared those antipathies—they'd all eventually relocated to New York, a backward journey of manifest destiny. But Abraham took to the desert. He made it his own.

Back in 1881, when Santa Fe's streets were still lit with turpentine torches, US Army captain John G. Bourke described its citizens as "a motley crew of hook-nosed Jews, blue-coated soldiers," and "señoritas wrapped to the eyes in rebosos." Outsiders may have seen Abraham as a damnable Jew. But insiders—the men who bought and sold and shaped Santa Fe—saw him as an ally. From early on, Abraham made powerful friends. He teamed up with the richest, best-connected men in Santa Fe: Thomas B. Catron, a local lawyer who became the state's first US senator, and for whom New Mexico's largest county is named; Stephen Elkins, New Mexico's territorial delegate and later President Benjamin Harrison's secretary of war; and a constellation of lesser luminaries, all pro-statehood Republicans. While Zadoc had been a Democrat, Abraham believed the future lay elsewhere.

Abraham ran for public office for the first time in 1879, for Santa Fe county commissioner, against Solomon Spiegelberg. Although Abraham outpolled Spiegelberg 834 votes to 821, the local elections board—

stacked with Democrats—ruled that Spiegelberg had won. The board did so by throwing out sixty-three votes in precinct number nine— populated by Hispanics, who tended to side with the Republicans— that were cast for "Abram Esstab." The case went to the territorial supreme court, which ruled that the votes should be counted. Thus, when the railroad arrived a year later, it was Abraham who drove in the silver spike.

Abraham was everywhere in Santa Fe's business. He was president of the gasworks. He was a director of the Texas, Santa Fe and Northern Railroad. He helped commission the construction of, and then won the contract to supply, the territorial penitentiary. The *New Mexican*—the local Republican paper—was a big booster. "A. Staab, Esq.," as the paper referred to him, was always "the most affable of gentlemen."

We can't know what happened between Abraham and Julia within the confines of the marriage—newspapers tell little of a man's inner life. We can't know whether he loved her, whether she loved him, whether he understood her melancholia or tried hard enough to help her. But we do know that Abraham was, without dispute, a genial public man. He was a natural operator. When the railroad to Denver opened, a reporter from the Denver paper, the *Rocky Mountain News*, joined Abraham and other business and army leaders in one of the inaugural train's private compartments. "They were telling stories, and there were some good ones told," the reporter wrote. "It is conceded that Major Hooper is the boss story-teller as well as the best, but Mr. Cornforth will not take a back seat for anyone else, not excepting Colonel Staab, of New Mexico, who can tell a story with a Hebraic flavor to it that is hard to equal." Abraham used his charm and his money, and later his "palatial residence," to strengthen his social and business ties. "Strains of music from the piano and violin . . . mingled with the sound of popping champagne corks, and time passed on golden wings," wrote the *New Mexican* about one railroad-wooing affair at the Staab house.

The local newspapers reported Abraham's comings and goings with near-tabloid voracity: "the merchant prince," they called him. He went to Las Cruces to acquire wool, and to Las Vegas to buy corn. He went to Socorro to raise funds for the territory's Republican convention. He went to Albuquerque to bid on county bonds offered at auction. He won, of course: "A. Staab, Esq, returned to-day from a successful business trip to Albuquerque," wrote the *New Mexican* in December 1888. "The qualification, however, is quite superfluous, for all of this pioneer merchant's trips are successful."

Successful in business, he certainly was; saintly, he wasn't. This was the West, after all, at a time when opportunities were wide open and the laws and their enforcers could rarely keep pace with changing conditions. Sister Blandina had written that Santa Fe was full of "land-grabbers, . . . quacks, professional deceivers . . . . I could use a half a dozen more adjectives and yet not touch on all the methods of deception carried on."

I could find no evidence that Abraham himself was a "professional deceiver," but it was clear that he knew how to exploit those gaps between what was and what should have been. He was a hungry, scrappy capitalist, and he didn't shy away from competition or conflict—though he preferred official, court-sanctioned justice to the frontier kind. Not a month went by without a newspaper reporting that Abraham had filed a lawsuit. He sued customers who hadn't paid their bills and local citizens who hadn't paid up on promissory notes; he sued the Spiegelbergs regularly; he sued business partners, friends, nephews, a brother-in-law; he sued the territorial tax collection authorities; he sued Western Union once, for failing to deliver a message at nine in the morning. The message had come instead at noon. If Abraham won, he collected. If he lost, he paid up. If he didn't like the legal reasoning behind his defeat, he set out to persuade the territorial legislature to pass laws more to his liking.

He wasn't the only Santa Fe player on the make: his colleagues Thomas Catron, who often served as Abraham's lawyer, and Stephen Elkins were every bit as grasping and litigious. The group was known as the "Santa Fe Ring." Catron was its brains; Elkins, Catron's brother-in-law, its grandiloquent figurehead; Abraham, one of its behind-the-scenes financiers. The editors of the *New Mexican* were also known to be members: hence the fawning press about Abraham and other members of the ring. They weren't an officially organized cabal, as far as history can tell—there were no officers, no minutes, no formal meetings—just a group of lawyers, politicians, judges, and business-men who socialized and played poker together regularly and kept the territory's business under their close purview.

The group had been implicated in unseemly deals since its inception in the 1870s. The *St. Louis Republican* newspaper had noted in 1876 that a "Republican ring" in New Mexico was, with small amounts of cash, able to induce the legislature "to do its bidding." In Colfax County, northeast of Santa Fe, Catron and Elkins presided over a company that gained rights to thirty-four Mexican land grants totaling nearly three million acres. The company was the largest landowner in New Mexico and among the largest in the nation. Abraham wasn't directly involved in the company, but he did loan money and invest in land in the county.

While Catron and Elkins kept their hands relatively clean, their company representatives in Colfax County played rough, acquiring the land by evicting Hispanic "squatters" from their ancestral lands and commissioning thugs to scare away or kill those who refused to leave. They sold worthless land that they didn't own to incoming homestead-ers, and foreclosed when the newcomers couldn't make enough money from the alkaline soils to pay the mortgage. They bought off govern-ment officials and hired outlaw gangs to rustle cattle and kill off their enemies. Abraham's business and political associates were part of those land grabs, even if Abraham wasn't directly implicated. "Nothing was too rotten," New Mexico's territorial governor Miguel Antonio Otero wrote in his memoirs, "for the well-known Santa Fe Ring to undertake." The ring knew how to game the system, because it owned the system.

The ring was also involved in the most famous New Mexico feud of the era—the Lincoln County War, in which Billy the Kid (the famous outlaw, not the lesser one of Sister Blandina's acquaintance) made his name in the late 1870s. The battle began as a conflict between ranchers competing for army beef contracts and expanded into full-scale bloodshed. Scrambling to protect their interests in government contracts in the area, Catron and the Santa Fe Ring were drawn into the battle. L. G. Murphy, who led one of the warring factions and is considered one of the "bad guys" of the conflict, happened to be Abraham's supplier of flour for the nearby Indian reservation and army fort. Abraham sided with him.

Ultimately, the US Department of Justice sent an investigator from New York, the attorney Frank Warner Angel, to figure out what was going on. In his notebooks, he tallied his assessments of the many New Mexico players he met on his travels. The list was arranged alphabetically, sort of. "Ayers, John," a federal Indian agent, he considered "Honest—Liquor his worst enemy." Axtell, S. B., the territorial governor at the time, was "Conceited—egotistical easily flattered Tool unwittedly [*sic*] of the ring—Goes off half cocked." And on it went up the alphabet, the notebooks revealing more about the investigator than the New Mexicans he attempted to assess. Angel, this man of the East, had his rules: people were "reliable" or "not reliable"; "honest" or "weak"; "shrewd" or "of no standing"; acting "for the right" or dreadfully wrong. Stephen Elkins was "Silver tongued—further comment unnecessary." The Spiegelbergs, who serviced the flour contract on the Mescalero Apache reservation in Lincoln County, he deemed "not reliable . . . use them against Z. Staab Bros & vice versa." Abraham, as the New Mexico face of "Staab Z & Bro.," had "Axes to grind," Angel wrote: "Not reliable—use them against Spiegelberg Bro & vice versa"—to what purpose, exactly, I don't know.

Abraham would probably have laughed if he'd seen Angel's notes, so earnest and naive. Because Abraham knew that "reliable" was beside the point in territorial New Mexico. Loyalty was the least of it. Bravado, charm, deception, ruthlessness, gunplay, grit: this was how you got things accomplished. Abraham and his Spiegelberg cousins played the territorial power game and played it well: Staabs with the Spiegelbergs, Staabs against the Spiegelbergs—it changed all the time. This was New Mexico.

nineteen

❧ PRINCESS GOLDENHAIR ❧

Flora Spiegelberg.

I hadn't known much—anything at all—about Abraham's Spiegelberg cousins until my first trip to the archives in Santa Fe. I hadn't known that they were the first German Jews to settle in New Mexico; I hadn't known that they had employed Abraham when he first arrived. I hadn't even known that they were relatives. But on the same day that I found the slim file containing Abraham's citizenship declaration, I discovered three much larger folders devoted to the Spiegelbergs. Flora Spiegelberg—the wife of the youngest Spiegelberg brother, Willi—had compiled them, and I found myself immersed in the blizzard of letters and articles and papers that told her story, and in the remarkable manner in which she seized the choices she was given in life and made them hers. It seemed that she had something to tell me.

Flora had arrived in Santa Fe in 1875, nine years after Julia, following a large wedding in Nuremberg and a yearlong trip in Europe. While Julia had spent the first days of her marriage on the Santa Fe Trail, Flora had stayed at the finest hotels and bathed in the finest spas. It was a honeymoon altogether different from Julia's.

Flora's appearance and temperament also stood in stark contrast to Julia's. Flora was tall, red-haired, and boundlessly peppy. She was, it was clear from her files, a *doer*—community oriented, a prolific writer and fervent memorializer of all things Spiegelberg. Her New Mexico exploits were celebrated in the *Jewish Spectator*, the *American Hebrew*, the *Jewish Historical Quarterly*, and a number of New Mexico historical journals. In her dotage—she died in 1943 at the age of eighty-seven—Flora wrote letters to librarians and archivists, dispatches from her New York apartment—heavy typeface, lots of typos. She was, said an interviewer from the *American Hebrew*, "a tall, straight old lady, with a most remarkable memory."

"I was born in New York in 1857," the *Jewish Spectator* article begins. Her mother had taken her back to Germany, however, after her father's death, and she met Willi Spiegelberg there in 1874—she was seventeen years old, he thirty. "I was young and he was handsome and in a very short time, I became Mrs. Willi Spiegelberg," she told her interviewer. After their European honeymoon the couple traveled from St. Louis "in very primitive steam cars" to the rail's end near Trinidad. The train had advanced swiftly in the nine years since Julia had traveled the trail, cutting off six hundred miles of prairie that Julia had had to travel by stagecoach. Still, Trinidad was rough. "The only hotel was a shabby, two-story building," Flora told the *American Hebrew*, and as she and Willi entered, "they found the enormous main hall filled with the tobacco smoke of a hundred cowboys," heavily armed and riled up after a roundup. When Flora entered the room, "probably the first of her sex that they had seen in a year," she speculated—"they arose as one man, swung their sombreros, and shouted with lusty enthusiasm, 'Hello, lady, sure glad to see you!' " Flora slept in her clothes that night.

The next morning, she rescued herself from a runaway stagecoach—slamming the door shut to avoid being thrown out when a train spooked the horses—then continued on to Santa Fe, eating chiles, beans, and buffalo tongue along the way. She did not relish the food, but she appreciated the novelty of it. She found the journey terrifying, but also considered it a wonderful adventure. She feared cowboys and Indians, but welcomed the sight of her first adobes. She miscarried along the way; this did not set her back, either. She seemed an altogether more durable woman than Julia, suited to such a life.

When Flora and Willi's stagecoach galloped into Santa Fe, a military band awaited them, playing Wagner's wedding march to welcome her to her new home. She was, she told the *American Hebrew*, the eighth "American" woman in Santa Fe, by which she meant non-Indian and non-Hispanic. In other reminiscences she declared herself the thirteenth. In others, the first.

Either way, she arrived in the desert with a splash. "Willi has a girl at last," noted the *New Mexican*. "She stands in the doorway of the handsome retail store of Spiegelberg Bros. to attract the eyes and arrest the attention of the passerby."

She placed herself quickly at the center of things. She threw parties that featured German cuisine and fine champagne; she collected art and founded literary and dramatic clubs, started a Jewish school, taught Jewish and Catholic Sunday classes, and created a children's playground in Santa Fe. It didn't matter that she was cultured and European and Jewish while Santa Fe was not. She did not require rich soil in order to flourish. She bloomed in the desert. The family built a lovely territorial-style home in 1880, two years before Julia's house was built. It was the first with running water and gas appliances, a sweet adobe blend of Europe and New Mexico, two stories with a pitched roof and territorial woodwork. Julia's home, just across the street, rose to the skyline, three stories of high French ornament and cultural affront. The two women seemed in every way opposites.

Flora was proud of her family's Western mettle: besides German, Spanish, and English, Willi spoke four Indian dialects, and he was, she said, an expert with the lariat and whip. He had traveled the plains fearlessly during the Indian Wars, Flora recalled, moving without an army escort and dodging throngs of "maddened redmen." Flora was equally stouthearted; by her account, she single-handedly talked down a lynch mob intent on dragooning her husband into hanging a pair of murderers, kept an eye on Billy the Kid while he shopped in her husband's store for a new cowboy outfit, and helped the author Lew Wallace, who was also New Mexico's territorial governor, polish his soon-to-be-blockbuster novel, *Ben-Hur*. She also claimed she was the conversation partner to whom General Sherman first uttered the words, "War is hell."

Flora, too, was close with the archbishop. He planted two willows in her front yard and they gardened together, speaking French. In her memoirs, she explained that the archbishop placed the Hebrew letters over the cathedral's arch in honor of his friendship not with Abraham, but with her own family.

As I read through Flora's files, I came to realize something: Flora was, as much as Julia—maybe more—the model for Paul Horgan's German bride. She played the piano beautifully, spoke perfect French, gardened with the archbishop, and entertained dignitaries. She was every bit the sparkling pioneer wife—comfortable, assured, adaptable—that Horgan described in his book. The German bride was a composite figure, I now remembered Horgan's saying in the book's introduction. The house in the illustration was Julia's, certainly. But many—perhaps most—of the stories were Flora's. Flora, not Julia, was the poised frontier woman in Horgan's book.

Even so, the frontier wasn't a place where Flora wished to live permanently. In 1893, she persuaded Willi to liquidate their assets and follow his brothers to New York, where her two daughters could live in a more cosmopolitan environment. In her new city, Flora was no less engaged. She organized the Boys Vocational Club and the Jewish Working Girls' Club, served on the Bill Board and the Daylight Saving Commission, and advocated relentlessly for the creation of a modern system of waste collection in the city (she was nicknamed "The Old Garbage Woman of New York" for her efforts).

Later, she wrote radio screenplays and a children's book, *Princess Goldenhair and the Wonderful Flower*. I managed to purchase it on the Internet: another artifact for my collection. It was whimsically illustrated in bright colors and with fine-lined detail, Yoda-eared dwarves and veiny-winged fairies dancing on mountainsides. My five-year-old daughter was fond of princesses, so I read it with her. The story involves a beautiful German redhead—similar in appearance, perhaps, to Flora?—named Princess Goldenhair. The young princess's troubles are fairly conventional—she is kidnapped by a wicked, poisonous-flower-wielding step-grandmother. But the rest of the book was hardly what we expected. After the step-grandmother drops the captive Princess Goldenhair off with some dwarves, she returns home to find her daughter dying in bed and her neighbor dead in the cellar with a broken neck. Then she lies down and dies, too. After three people perished in three pages, my daughter's eyes fogged with tears, and I decided that

Flora was probably a better fabulist of her own charmed life than of fictional princesses. We stopped reading.

Flora was also a peace activist: in 1919, after the First World War ended, she wrote a petition titled "The Ten Commandments for World Peace," which demanded an amendment to the US Constitution requiring that all future wars be decided by popular referendum, and asked that all "national, racial, and religious hatreds should be eliminated." She was tireless in her advocacy, mailing her petition to newspapers around the country and to the Library of Congress, and persuading the Veterans of Foreign Wars to discuss her petition at their annual meeting.

Even in her dotage, Flora found idleness unendurable. During the last thirteen years of her life, she engaged in exhaustive correspondence with New Mexico museums, historians, and archivists, building a record of the Spiegelbergs' role in New Mexico and her own as New Mexico's best-known German bride: "I presume I am the oldest living American Pioneer Woman of Santa Fe," she wrote to one archivist. "Dear Friend, kindly pardon all corrections, for unfortunately I have but ONE EYE," she wrote to another when submitting a chronicle. Her *New York Herald Tribune* obituary ran with the subhead "Once Ate Buffalo Meat." I found myself both admiring her self-regard and fortitude, and also finding her slightly unbearable. I was jealous on Julia's behalf.

It was as if Flora, born into the waning half of the nineteenth century, inhabited not only an entirely different generation, but a different world from Julia's. They were conceived only thirteen years apart, but modernity itself seemed to separate them. Victorian languor was going out of style; women began to flee the fusty parlor. There are no reports of ghosts in Flora's house; she was too busy with the present to linger in the past.

Whereas Julia led a quieter life. Besides her seven surviving children and the distinguished brick home, she left little record of her passage through thirty years in Santa Fe. Julia didn't see herself as a pioneer or a Western heroine; she was not one to love the view from such vast distances. In the newspapers, she always appeared as "Mrs. A. Staab." She was an adjunct and helpmeet to her husband, cloistered and Victorian, a creature of both her waning epoch and her own reticent and melancholy constitution. Were it not for the stories they tell of her ghost, we'd know nothing of her life at all.

And yet she is now New Mexico's most famous German bride—and the Staab we care about most.

twenty

❧ BOODLE AND PAYOLA ❧

The Z. Staab Brothers building in Santa Fe.

Back in the waning years of the nineteenth century, though, Abraham was the name on everyone's lips. In the years before his trips with Bertha to California and with Julia to Europe, he was at the very top of Santa Fe's social and political pyramid. In January 1890—a year before the family headed off to Germany for Julia's cure—he at last changed the name of his firm from "Z. Staab & Bro."

to "A. Staab."—he was his own man now. Abraham was fifty years old, Julia forty-five. Their family was, by all appearances, thriving. The children—Paul excepted—were healthy. They were marrying, courting, heading off to university. Julia was still making the social rounds, out in the world. Everything was right. If only the world could have stopped revolving.

In November of that year, after a few years' hiatus, Abraham again ran for county commissioner. The election was contentious; the local Democrats were not fond of Abraham. In the previous election cycle, the *New Mexican* reported, the talk at the Democratic county convention "consisted mainly of abuse of Mr. A. Staab." This time, if anything, the malice had grown. The Democrats accused Abraham of pocketing money from county bonds issued a decade earlier to entice the railroads. The allegations rose to such a drumbeat that Abraham felt forced to reply. "The insinuation in the *Santa Fe Sun*"—the Democratic newspaper—"that I had any part in, or knowledge of, any '$16,000 bond steal of years ago,' is willfully and maliciously false and unfounded. A. Staab," he signed it. The *New Mexican*—the Republican newspaper—in turn accused the Democrats in power of embezzling county funds. "They hate an honest man with a deadly hatred," opined the *New Mexican*. "They know that if Mr. Staab is elected, their career of crime and corruption is at an end, and that if there is any law in this country they will be punished."

And so it went.

When the votes were tallied on election day, Abraham was found to have tied his opponent, a fellow dry goods merchant named Charles M. Creamer. The Democratic election judges, however, had thrown out a dozen votes for Abraham because they had been written in lead pencil instead of ink. To resolve the tie, the two candidates' names were put in a hat, and Abraham's was chosen.

All hell broke loose. The Democratic county clerk, insisting that both of the names in the hat had been Abraham's, refused to certify the election. The previous Democratic county commissioners also refused to step down. In order to take office, Abraham and his fellow Republicans formed a "shadow commission" and elected Abraham chairman. They must have had the sheriff on their side, because they managed to send the Democratic commissioners and county clerk to jail for contempt of court. The contested election eventually made its way up to the US Supreme Court, which ruled on Abraham's behalf.

At the same time, the Democrats were in a lather about militia warrants. These were IOUs that funded armed squads the territory had raised to fight the Indian Wars in the late 1860s—money promised to volunteers at a later date if they joined the militias. After the battles subsided, however, the territory had no money to reimburse the volunteers, and the federal government had no interest in doing so, either. A few years later, Abraham, Thomas Catron, and some of their associates bought hundreds of such warrants for pennies on the dollar, hoping to cash them in later for dollars on the dollar. To persuade the legislature to recognize the warrants, many of which were probably fraudulent, Abraham offered, in a letter someone turned over to the *Santa Fe Sun*, $150,000 in payments to the territory's legislators—$5,000 per legislator—to "defray the expenses" of their work in legalizing the warrants. Democrats believed it was a "bribing fund"—"a gigantic scheme," the *Sun* wrote, "to debauch the legislature."

The legislature formed a special committee to look into the militia warrants and the "Staab letter," as it came to be called. The committee asked Abraham to testify about his efforts to sway the legislature, but he said that he was sick and could not give testimony. Abraham may not have been "the Al Capone of Santa Fe," as the ghost stories suggested, but there was indeed a whiff of corruption about Julia's husband. In early January 1891, the *Sun* reported that Abraham "has been recommended by his physician to go to sea level as soon as possible." Abraham ignored the advice at first. He recovered inside his Palace Avenue home, while the *Sun* continued to accuse him of "payola schemes," of redirecting campaign money, and of general "boodle." In mid-February, Abraham's health grew worse. "A. Staab is quite ill again," reported the *Deming Headlight*, "and will, in all probability, leave for California during the coming week, to consult a specialist upon his ailment."

This time, Abraham did leave, repairing to Los Angeles, Redondo Beach, and San Diego. This was the same trip on which he and Bertha went wine tasting and saw the Great Pacific, on which Bertha duked it out with the Miss Pullmans for the meager supply of young gentlemen. Though Bertha had mentioned that Abraham was ill at times, he had seemed well enough to enjoy their travels. Yet while in California, he resigned his spot on the board of commissioners, citing his fragile health.

The papers fell quiet until early April, when the *Albuquerque Democrat*, keeping tabs on Abraham in California, brought forward a scandalous charge: "A. Staab, of Santa Fe, who is in California with his daughter on a health-seeking trip, recently lost $30,000, playing poker at Redondo Beach; and after giving his checks for that amount, he telegraphed home to stop payment on them." Thirty thousand dollars—1891 dollars—in one night.

Aunt Lizzie wrote in her family history that every time Abraham played poker, he left a twenty-dollar gold coin under all his kids' pillows. She believed it was to persuade them—and Julia—that he always won. Julia's husband was a gambler, in every way. That's how you made it in America. He had gambled on crossing the ocean at the age of fifteen; he had wagered on the hazards of transporting people and goods on the Santa Fe Trail; he had ventured to import a wife he barely knew; he had speculated in real estate and irrigation schemes and mines and militia warrants. Likewise, he gambled at the table.

But he didn't always win. He lost thirty thousand dollars in that one sitting in Redondo Beach. He had lost the army headquarters, too, and the lucrative contracts they provided. He had selected a wife unsuited to the intrepid life he had chosen. And he had wagered that when New Mexico was admitted to the union, the militia warrants he had purchased would be paid off by the legislature. If they were, he stood to make more than a million dollars on the redemption.

The warrants never paid, however. I found some of them, yellowed and forgotten, in an envelope in Abraham's slim folder in the New Mexico state archives, crowded alongside Flora Spiegelberg's more expansive reminiscences. The warrants meant nothing to me when I first discovered them; only after reading the newspaper accounts of Abraham's scandalous efforts to collect those IOUs did I understand just how much they meant to him. They obsessed and plagued him, these promises of easy riches—and they almost undid him.

Abraham had always been a careful curator of his reputation. Now, for the first time, he was a figure of ridicule. In every life, lucky streaks end. Lives go from golden to cursed, or merely to ordinary. The tide turns; the gods become mortal. "A. Staab, the Santa Fe merchant, whose speculations in New Mexico militia warrants and poker games, have become common talk in the territory, passed through today journeying home from the west," wrote the *Las Vegas Optic* as he returned from California with Bertha in April 1891.

No wonder Abraham and the family turned right around and headed off to Europe. The Continent offered one kind of cure for Julia, and another for Abraham. Only one of them would be successful.

Margaret

◇◇

"I LOOK FOR GHOSTS; but none will force their way to me," wrote William Wordsworth about a woman named Margaret, who had lost a child and been undone by grief. "The very shadows of the clouds / Have power to shake me as they pass." The Margaret of the poem felt a terrible urgency to find her lost child's ghost. My search for Julia was less fraught, but like Margaret, I was trying, through various assorted methods, to speak to the dead.

I decided to give medical marijuana a try. The psychics I consulted had told me that Julia might address me directly, if only I was open to it. It was the openness that vexed me. I am not a regular marijuana user—it makes me anxious. But a friend with a medical prescription gave me a cookie, and I took a tiny bite. Nothing much happened at first; I fell asleep. And then I woke an hour later, and I realized that everything looked pink.

This was OK, initially. But soon my thoughts began to tiptoe out of my head before I could finish thinking them. And my heart began to beat very loudly, and I began to worry for my health, for my future, for the world. My husband lay beside me in bed, sound asleep. I contemplated waking him, but I couldn't imagine what he could do to help me. I glanced at my bedside clock. It said 11:11. An hour later, I looked again. It said 11:13.

I went downstairs and threw up. My stomach hurt not from gastrointestinal complaint but rather from panic. I wanted the cookie to go away. I brushed my teeth for what seemed an eternity, then turned on my phone and pulled up a search engine. "How long does edible marijuana last?" Though I knew the answer already: many more hours. I threw up again, commenced another epic tooth-brushing session, and tried another search:

"How come down weed?" I read something about orange juice. I drank orange juice. I tried to read, but I couldn't make sense of the words. I turned on the television; I couldn't follow the plot. I sat bolt upright on the couch in the kids' playroom, staring straight ahead. Back to my phone: "Die marijuana overdose?" I was only slightly reassured by the answer.

How had it been for Julia to battle these interior demons, the ghosts of the mind? My problem was now inside me, as Julia's must have been. I thought of Emily Dickinson—another shut-in: "One need not be a chamber to be haunted," she wrote. "One need not be a house; / The brain has corridors surpassing / Material place." Better to meet an "external ghost," she wrote, "Than, moonless, one's own self encounter / In lonesome place." I didn't want to meet myself that way. I didn't want to be haunted.

I sat wretched, hugging my knees on the playroom couch, and I knew that if ever there were a time when I might be receptive to a voice from the other side, it would be now—my brain was unpeeled, my rational mind collapsed. I was open. I called for Julia. I waited with my hands on my thighs. I closed my eyes. I implored her: Come. Tell me what happened. Speak of the past.

But she didn't, not yet.

twenty-one

❧ TALE OF WOE ❧

Julia's mother, Henriette Schuster.

By the time Abraham returned to Santa Fe from Europe, the papers seemed to have forgotten, or forgiven, his scandal-tinged departure—the ugly election, the gambling losses, the militia warrants. The *Santa Fe Sun*, which had goaded him ceaselessly six months before, welcomed him back: "A. Staab, one of the oldest and richest wholesale merchants in the territory, has returned to his beautiful

Santa Fe home from an extended European tour." The *New Mexican* commented on his improved health. "Hon. A. Staab returned Saturday night looking hale and hearty and ready for any amount of business." A few days after Abraham's return, he sued his nephew Alexander Gusdorf for $29,861.85—for what, I couldn't ascertain. Abraham was, indeed, ready for business.

Julia, meanwhile, had moved into a private villa at the spa in the Harz Mountains. Bertha and Delia accompanied her, tasked by Abraham with keeping her close and meeting her needs. "Tues 25 Aug 1891—We have unpacked everything," Bertha wrote.

The villa is built very prettily; it is quite large and is situated right in front of the woods on a grassy slope overlooking the town. The air is very fresh; we can actually feel that it is pure. The principal thing is the utter restfulness of the place; it is very quiet—quieter even than Santa Fe. Astonishing but true. If mamma's nerves don't get strong here, I don't know where they can.

Things were indeed quiet at the spa. For Bertha, young and impatient, time seemed to sputter out. "It seems ages that we've been here," she wrote the day after they arrived.

We embroider, walk and read and read and embroider and walk— don't know a soul and not likely to make acquaintances. Few people in the house and they are mostly old and have their meals served in their own rooms—but if mamma only gets well quickly all will be well.

In the days that followed, the girls continued to embroider, walk, and read. They composed witty poems directed at various young gentlemen of their acquaintance ("you pass the girls without a smile / and act in altogether a shabby style"). Time dawdled, excruciating and unhurried. "Monday we will be here three weeks," Bertha wrote in mid-September.

As yet mamma is not better. Her spells take place about every other day; she has one good and one bad day. To-day she has had the worst nervous attack she has ever had during this illness—hope it will not return again—Mamma was so discouraged that she wanted to go away from here at once—of course we cannot do that, for where shall we go?

Not knowing what to do, they wrote their aunt Adelheid and asked her to stay "for a week or ten days in order to help encourage mamma and so that she has somebody with her, who is more experienced than we are." Julia saw her doctor each day. Sometimes he'd call on her, other times she'd go to see him. "He says mamma will get entirely better, but that nervousness doesn't go away in a day but requires time and courage—The patient can do more for herself than others can do, by trying to be calm."

A Mrs. Flechtheim—another spa patron—visited from time to time "to talk and enliven mamma." Bertha and Delia would take walks with the Flechtheim family, "changing off" so that Julia was never left alone. They did a bit of sightseeing when they could, visiting whatever high points and monuments they could find in the nearby hills. In the absence of young gentlemen, Bertha flirted instead with Julia's doctor.

I have made a conquest of him. He is old and married so there's no danger! He told me that he had always imagined 'Queen Mab' in "A Midsummer Night's Dream" just like my sweet self!

Childish fancies, they were, but more enjoyable than what Bertha beheld in the earthbound world—the frightening tenacity of Julia's illness, the tide of her health ebbing further from shore. Julia was sick—in her head, and in her heart. "I hope this won't last much longer, for it is dreadful," Bertha wrote.

When Tante Adelheid arrived, Julia improved for a time. Adelheid was a "splendid nurse," and everyone quickly cheered up. "Her arrival had a very good effect upon mamma's condition," Bertha wrote.

> *The day before aunt came Delia and I almost despaired, mamma looked so strangely and talked so mournfully. It seems that her condition was such a critical one that the least excitement, bad or good, could turn the scales for better or worse. Aunt's arrival accomplished that; the beautiful change brought about by her coming seemed to take place at once and a slight but steady improvement has been going on ever since. Thank goodness the worst days are over and we need have no fears for her eventual entire recovery.*

Bertha was happy to think that in her letters to Abraham, she might be able to offer more encouraging news, not the "same old tale of woe, 'no change for the better.' Now at least we can afford ourselves and him the satisfaction of hoping for speedy recovery."

But Julia's improvement was short-lived. After Adelheid went home, Julia returned to bed, and the family determined that the Harzburg spa had healed her no better than any other. They would stay another week in "this dullness and quiet," Bertha wrote, and return to Hanover to be near Julia's mother and sisters. "There is absolutely nothing new," Bertha wrote. On sad days, Bertha reminded herself, "This too will pass away"; on gladder days, she tried to "enjoy the present."

They left Harzburg in early October. "Mrs. A. Staab and daughters are in Hanover, Germany," the *New Mexican* reported. "Mrs. Staab has been quite ill, but is slowly recovering and gaining strength." But the *New Mexican* had it wrong. Julia wasn't improving. She was descending into an underworld from which the sunlit earth seemed more and more remote. Abraham, from afar, and his girls, at Julia's side, could only look on as winter fell.

After a few days in a Hanover hotel, the three Staab women found longer-term quarters for the winter. "Rather small rooms but comfortable and what is most important mamma can have (in the way of eating) what she wants and that's at any time."

On November 13, Julia's mother, Jette, the widow Schuster, died. The family held a service two days after her death, but didn't tell Julia for another five days. "She had a very severe crying spell but after we had calmed her no bad effects," Bertha wrote, "her nervous attacks are shorter but not less frequent yet."

After that entry, Bertha wrote in her diary less frequently. In the weeks that followed, she and Delia had their photographs taken. They watched over their mother each day. Bertha was seized with a fit of self-loathing. "I wish I were beautiful in every way, rich, charming, lovely, happy!!" Winter settled in. Delia wrote to the Santa Fe newspaper, reporting that Julia was slowly improving but that the weather in Hanover was "simply abominable."

They celebrated Christmas. Bertha didn't seem to think this unusual, or if she did she didn't mention it—this is what German Jews did in those years. If they were religious at all, these assimilated Jews tended to practice a mild "reform" version of their faith, couched in the language of the Enlightenment. They worshipped in gilded, neo-Gothic synagogues difficult to distinguish from the nearby churches, with pipe organs and German prayer books. They ate pork—bregenwurst, bratwurst, liverwurst. They spoke German exclusively—no more Yiddish. And on Christmas, they adorned fir trees with scented wax candles.

That Christmas in Hanover, Bertha received a case for her spoons from Delia, a black lace Spanish shawl and a dozen embroidered handkerchiefs from Julia, a small painting from her brother Paul, and a new ring from Abraham, to replace the one she had lost in Los Angeles. She and Delia gave Julia a silver tea sieve.

On New Year's Eve, they went to bed well before midnight—Bertha had come to expect less festivity in her life. During the short days of January, they learned to ice-skate, "and when we had learned, it thawed and thew and watered and sent our skates to rest on a nail in the wall." And then, in mid-February, an awful thing befell Julia—something so bad that Bertha couldn't bring herself to write of it. "Mamma had been progressing very very slowly from Nov till February 12," Bertha wrote.

> So slowly, that sometimes we doubted whether she improved at all. The last few days before that dreadful accident, we thought everything was going along nicely.—Delia thought of going to Cologne at the end of the month and our plans were beginning to look real and the future seemed to have a tinge of rosiness.
>
> But no—that awful day the 12th—I shan't say anything about it—everything has turned out well and we must be thankful that it is not worse.
>
> Mamma is in bed, but will get up in a few days. The bandages are to be taken off Friday and we pray and hope all will be well—
>
> A letter was sent to Papa—instructing him and telling him all facts. The doctor said it was our duty. We have a sister to attend to Mamma's wants.

And there the diary ended, along with the family's hopes for a happier result for Julia. I imagined that the days and weeks that followed the abrupt ending of the diary must have been terrible for Bertha and Delia. For me, though separated from those raw emotions by the large gulf of history, it was an unhappy leave-taking, as well. I'd learn no more about Julia's "dreadful accident." Bertha, who went on about every "young gentleman" who crossed her path, couldn't bear to explain what had happened to her mother. Julia had been doing better, she said—which suggests that Julia's condition took a turn for the worse, and that the "accident" somehow involved her condition. She was in bandages. It was months, now, after the surgery on her womb, so it was unlikely that the bandaging had anything to do with her surgical site. Had Julia wandered in the night and fallen? Had she been cut somehow? Wounded herself in the bathtub? I couldn't know. Bertha wouldn't tell. It was too horrible.

Abraham and the girls had trusted that Germany would help. Superior doctors, unrivalled spas, familiarity, family, sisters, home: they'd believed in happy endings—in an American fairy tale, all triumph and no decline. But Germany was no panacea. Bertha's grand European adventure, Julia's quest for health and healing—it was all ruined.

twenty-two

❧ THE ANGEL OF NEUHAUS ❧

Wolfgang Mueller as a young man.

I didn't want the diary to end. There was so much I still needed Bertha to tell me. The last page contained a scrawled ledger noting what she had paid for various items: books, stamps, a curling iron, coffee, and three spoons (two dollars each). I scrutinized it, looking for answers.

I was disappointed that Bertha hadn't told me everything I needed to know. But I was also grateful that something so personal had survived the long years at all. It was a gift. The tactile realness of the document made me feel that much closer to my troubled family, and to the past—filtered though it was through Bertha's twenty-one-year-old eyes, and stifled by all the taboos of the era: sex, female problems, illness, madness. I would have to accept that Julia remained unknowable, by dint of both time and disposition. But I now knew something, at least, of her moods and her travels, her pain, and the struggles of her children and husband as they tried to save Julia and came to grips, by stages, with the inevitability of her decline.

That was more than most of us know of our forebears. In most families, we have only stories, told from parent to child and from those children to their own, handed from generation to generation, stretched and twisted and muddled with each new telling.

I would have to rely on hearsay from there on out. I was of course a generation or three too far removed to learn anything firsthand. When my grandfather died in 2007, he had been the last of Julia's living grandchildren—and he was born more than a decade after Julia died. Now, our oldest living relative was a woman named Betty Mae Hartman, my father's second cousin and Julia's great-granddaughter. She lived alone in a small adobe in downtown Albuquerque that was nothing like the stately home of her great-grandfather. It was a one-story building of modest ambition. Apparently, the dry goods dollars hadn't sustained this branch of the family any more than they had my own. I visited Betty Mae—meeting her for the first time—on a blustery spring morning in 2012. She was ninety-three and walked unsteadily, ushering me through a dark carpeted living room full of family photographs to a large sunroom where she spent much of her time in a comfortable upholstered recliner. The room was cheerful and bright, cluttered with crocheted blankets and cushions and doodads.

Betty Mae had once been a beauty—a dark and dainty, simmering Jewish beauty. She still had the lush black eyelashes of a younger woman and razor-sharp cheekbones, a scaffolding of loveliness that had withstood the havoc of age. Anna, Julia's eldest, was Betty Mae's grandmother. When Betty Mae was young, Anna lived with her husband, Louis, in a grand European-style home that Abraham had built for them in downtown Albuquerque; it had two stories, two parlors, a solarium, servants' quarters, china service for eighteen, exotic birds, tennis courts, a garage, and two barns—one for horses, one for cows. In the summertime, Anna and Louis packed up the cows and moved to Abraham's house in Santa Fe, where it was cooler.

Bertha, married by then to my great-grandfather, Max Nordhaus, lived down the street from Anna and Louis in an elegant neoclassical building. The two sisters were close—thick as thieves, Betty Mae told me. She perched in her chair in the sun, sharing stories from her childhood in Albuquerque, and when the memories ran out, we looked at photos from those long-ago times.

Before she sent me on my way, Betty Mae mentioned that she had, as a teenager, "gone out" with a second cousin named Wolfgang Mueller, who was a refugee fresh from Germany. He was, she told me, a grandson of Julia's youngest sister, Emilie. He knew all about Julia's family in Germany, she said. I found his phone number, and when I returned home to Colorado, I called him.

Wolfgang lived in Washington, DC, where he had moved after serving in the army during World War II. He had run a meat supply company for many years, and then, in his seventies, he'd started a fish supply company. "If you visit," he promised, his voice resounding, German accent still thick, "I will feed you fish. Very fresh." I mentioned that I had met Betty Mae. He had also seen her recently in New Mexico, he told me. I started to tell him that she was still quite striking, but he interrupted. "At one time she was very lovely," he said, "but she has aged a lot." He paused for a moment. "I have weathered the time very well."

He had indeed. At the age of ninety-three, he still played tennis almost every day. He traveled to Europe regularly—he was leaving in a few weeks, he told me, for a river cruise on the Danube. It took me until autumn to arrange to visit him at his large and tastefully appointed townhome in Washington, where he did, as promised, feed me very fresh halibut. He looked as if he were still in his seventies, with a full head of brown-streaked hair; his eyebrows were black. He had a robust bearing, and his plummy, deep voice contained none of the quavers of age. His skin was smooth, preternaturally so—I wondered, idly, if he'd had some work done. In this family, there were people like Wolfgang, with his booming confidence, and like Flora Spiegelberg, with her civic projects and campaign for world peace, and my grandfather, who fought World War II on skis and built a tram up to the crest of the Sandia Mountains and spent a long legal career fighting for Native American rights. These were people who lived long and full lives. They were satisfied, immodest, unapologetic: happy.

Wolfgang didn't remember Julia; he had lived almost a century, but he was still too young. He remembered some of Julia's sisters, though. They were, he told me, a "very distinguished Jewish family in Westphalia." Their children studied at the best schools and universities, joined tennis clubs, collected art. The Schusters broke barriers: children of merchants and peddlers, they were elected to the Reichstag and became leaders in the banking sector, in the law, and in publishing. America wasn't the only place where a Jew could make inroads into the larger world. Julia's nephew, Arthur Nussbaum, son of the irksome Uncle Bernhard who had so annoyed Bertha in Bad Pyrmont, became a renowned expert in international law and eventually moved to New York to teach at Columbia University. Two other nephews, Wolfgang's uncles Heinrich and Charles, served in a hussar regiment during World War I. Charles became a commissioned cavalry officer; a Jew could not have dreamed of such a thing a generation before. It was a short-lived dream, of course—and Wolfgang's family would come to understand that intimately.

Wolfgang remembered his grandmother, Emilie, well. She was the last-born of her siblings, seventeen years younger than Julia. Neuhaus, the small town where she lived, lay only twenty minutes by streetcar from the city of Paderborn, where Wolfgang grew up, and he saw her often. At age six, he went to her home to recover from pneumonia and stayed for a full year, with Catholic nurses watching him around the clock. Emilie was nervous and domineering, Wolfgang said. He suspected that these traits were common to Schuster women, Julia included. "They had a little bit loose screw," he said, twirling thick fingers at his temple. "My grandmother was not easy to get along with. I was supposed to love my grandmother but I never liked her that much."

In her younger years, Wolfgang said, Emilie had been "extremely good-looking. Magnificent!" She was also very wealthy. Emilie's husband, Louis Rosenthal, owned a six-story flour mill whose grinding wheels spun on the river Alme, which ran through the center of Neuhaus, a pretty town with winding streets of *Fachwerk* homes and a moated, white-baroque castle fortress. The family's home, on the main road leading into the village, was a mansion in the art nouveau *Jugendstil* manner, with a tremendous yard, a tennis court, rosebushes, hazelnuts. Emilie donated generously to the local church and gave dresses to the town's young girls for their Catholic confirmations. "They called her the Angel of Neuhaus," Wolfgang told me.

Louis died of a heart attack in 1912 while swimming in the North Sea, leaving Emilie to raise their teenage children alone—three boys, three girls. She hired a manager to help her run the mills, and ran them ably until 1933, when she was seventy-two and the Nazis came to power. First they took the mills. Next, her home. Then she was told to pack a bag. She packed carefully: tailored dresses, starched undergarments. "And they marched her down the street in Paderborn, this old woman," Wolfgang said, "and they put her on a train, and took her out of town."

◇

I had always assumed that my family had lost relatives to the Holocaust—distant ones. But I had never known names. Now I did. Wolfgang gave me a detailed Schuster family tree. Below the names of those who had perished, someone had typed, "Died in concentr. camp." Most of Julia's siblings had been lucky enough to be born too early; they didn't live long enough to die in a "concentr. camp." Emilie did, however.

Wolfgang escaped such a fate, and my family in New Mexico is one reason why. Wolfgang was thirteen in 1932 when the Nazis won huge gains in the German parliament; his father said not to worry, it was just politics. "My father was more German than the pope was Catholic," Wolfgang said. Wolfgang hadn't thought of himself as particularly Jewish until the Nazis took over. Then friends were arrested and taken away, and he grew afraid to go to the local swimming pool for fear others would see that he was circumcised. "I hated that I was Jewish," Wolfgang said. "You were like an insect." When a teacher entered his classroom after Hitler was appointed chancellor in 1933, the students were required to say "Heil Hitler." Wolfgang refused. The principal knew Wolfgang's father, and told him to send the boy out of the country.

Wolfgang's parents quickly enrolled him in a boarding school in England. And this is where Julia's New Mexico family came in. When it became impossible to send money out of Germany to pay the tuition, Wolfgang's parents asked Bertha's husband, Max—Bertha had died a few years before—to pay Wolfgang's tuition in England. Max did so willingly; he had already brought over or sponsored a number of German relatives now living as refugees in America. After a few months of supporting Wolfgang in England, Max offered to bring him to America to work—he was in the mercantile business, as Abraham had been. For Wolfgang, it was more than a rescue. He was a voracious reader of the German Western novels of Karl May, and New Mexico was the place of his *Cowboy und Indianer* dreams.

In 1936, at age sixteen, Wolfgang boarded a boat to New York. The journey was a grand adventure for him—he had no idea, as he steamed to America, what was in store for his relatives in Germany. He enjoyed the trip immensely. On the ship, he reigned as the boat's Ping-Pong champion and "befriended a young lady." In New York, he took in the sights, ate a banana split at an open-air ice cream stand, and went to Minsky's to watch a burlesque show. Finally, he boarded a train across the plains to Albuquerque.

He arrived to bad news. Max had died only days earlier—a heart attack, at his desk. Max's children—my grandfather and great-aunts Lizzie and Maxine—knew nothing about Max's plans to hire Wolfgang. But they took him in gladly. Accustomed to wartime scarcity, Wolfgang was amazed by the "opulence" of his first breakfast: pastries, juice, grapefruit. The family set him to work, giving him no special dispensation except an occasional invitation to a family dinner, and paying him fifteen dollars every two weeks. To start, he packed piñon nuts to be sold in the East. On the bags he wrote, "Packed by Wolf, a Zuni Indian."

Two years later, in the fall of 1938, Wolfgang's mother, Anna, visited him in Albuquerque. She had been there only a short time when his father sent a telegram telling her to delay her return. Kristallnacht had intervened—the "night of broken glass," of arrests and beatings, of burned and looted Jewish shops and homes and synagogues—and it was too dangerous for her to go back to Germany. Anna possessed only a visitor's visa, however, and these were no longer the days when the Santa Fe Ring ruled the state; no number of pulled strings could change Anna's documents. She couldn't legally stay in Albuquerque.

But Mexico was not far away. So Anna moved to El Paso, where her uncle Bernhard—Emilie and Julia's brother—had settled after leaving Santa Fe in the 1880s. She obtained an address across the Rio Grande in Juárez, Mexico, and each month she'd walk from Bernhard's house in El Paso across the bridge to Mexico to renew her immigration request.

Wolfgang had never met his uncle Bernhard before. Bernhard had once had a mercantile business in El Paso and also owned a 115,000-acre ranch across the border in Mexico, where the family had farmed and made wine. Now he was in the insurance business. He was quite old, close to eighty, and short, like all the Schusters and Staabs. He wore a sombrero, had a waxed mustache and a perpetual cigar in his mouth, and spoke perfect Spanish. Bernhard's wife, also named Emilie, was half Mexican, half Jewish. "She was like a queen," Wolfgang said, imperious and regal. Their only daughter, born in 1900—four years after Julia's death—was named after her aunt: Julia.

Wolfgang's mother lived in the basement of Bernhard's home and gave manicures part-time in a beauty parlor. When Wolfgang visited on weekends, he and Anna dined with Bernhard and his wife. They served chili and beans, Wolfgang said, and at dinner, "there was not much conversation." Anna stayed with her uncle Bernhard until she was able to secure a visa, then moved to Washington, DC, to be with a daughter who had moved there.

Wolfgang was drafted into the army soon after—he served as a translator in the Intelligence Corps, interrogating German prisoners. Wolfgang's father, Ernst, a successful lawyer before the Nazis took power, escaped to London after he learned he was about to be arrested, borrowing money from cousins in Paris to make the trip; those cousins were later killed at Auschwitz. In London, he made his first pennies carrying furniture out of houses bombed by the Nazis. He met a woman in London who made him happy, and persuaded Anna to give him a divorce. She later regretted it. She and Ernst had lost all they had: their money; their home; their community; their marriage.

They were, of course, the lucky ones.

twenty-three

❧ OTHER MOMENTS CONTRIBUTE ❧

Emilie Schuster Rosenthal with her daughter, Hilda.

Julia's sister Emilie was not one of the lucky ones. She was wealthy. She was strong of mind and constitution. She had been generous to the people of Neuhaus and Paderborn. But none of these things could save her in the end.

I visited Neuhaus and Paderborn on my trip to Lügde with my mother. A local historian named Margit Naarmann took us around.

She was tall and elegant, with big, round features and plush cheekbones. She wore a hairband in her dark hair, and silver button earrings. Margit was not Jewish. She had grown up in a small village in the countryside, knowing nothing as a child about what had befallen Germany's Jews. But as a young woman, she had worked in Scotland as an au pair for a German Jewish family that had emigrated during the war, and she was horrified by their stories. She'd returned to Germany determined to educate others about what had happened. She taught at the University of Paderborn, and she wrote books about Paderborn's Jews, including one that told the story of Emilie and her family.

Paderborn was a well-tended city—cobbled streets, old baroque buildings, new buildings that looked like old buildings, and striking modern buildings, too. The Pader River welled from under the city's heart, thousands of underground springs spreading tendrils and branches throughout the city. The people of Paderborn were tall, I discovered. My mother and I, five foot three and five foot five, respectively, felt lilliputian by comparison—how the Schusters must have stood out! The Paderborners were quite pleasant, if reserved. It seemed an altogether moderate place.

But of course Jews fared no better there than elsewhere in Germany. It was no Nazi stronghold—the city voted for the Catholic party, Margit told me. Nor had Paderborn's citizens initially supported the Nazis' restrictions against Jews. Julia and Emilie's sister Amalie, for instance, had married into a family that owned the largest department store on Paderborn's main square. When the Nazis ordered a boycott of Jewish businesses, many Paderborners refused and continued to shop at the store, until party thugs blocked off the entrances and made it impossible to do so. Amalie died of natural causes soon after: in retrospect, hers was the kinder fate. Her grandson Karl Theo, who ran the store, was sentenced to a nearby prison in 1938 for the crime of "racial transgression"; he sold the department store at a fraction of its value. After his release in 1940 he escaped to Palestine, where he worked as a chauffeur.

Emilie's mills in nearby Neuhaus were similarly "Aryanized." Her partner in running the company, Carl Schupmann, wasn't Jewish, so in 1933 they changed the name of the company from "Rosenthal and Schupmann" to "Schupmann and Rosenthal." In 1935, they changed it again, to "Schupmann and Co." That wasn't enough: they sold the company in 1937—the rye, wheat, and threshing mills; the office, mill-wright house, pond garden, and meadows along the river Alme—to three businessmen who had no troublesome Jewish associations. A year later, on Kristallnacht, Emilie's son Arnold was arrested and briefly imprisoned, bused past a mob of screaming Hitler Youth and Paderborn's burning synagogue. In September 1939, Emilie was forced to move to a *Judenhaus*—one of five buildings crammed with Paderborn's remaining Jews. Her sons, Heinrich and Arnold, and Arnold's wife, Hilde, were sent to the same house.

Emilie tried to leave. She was almost eighty years old. Cuba was still accepting Jewish refugees, but the visas cost money. So did the draconian "departure taxes" required by the Nazi government, and Emilie had no more money. In the United States, her daughter Anna—Wolfgang's mother—worked frantically to make arrangements. "We are all very happy that mother has finally agreed, so to speak, to make plans to travel to Cuba," wrote Emilie's son Arnold in the fall of 1941 in a letter to Anna, reprinted in Naarmann's book. Arnold and his wife, Hilde, hoped to emigrate with Emilie; another brother, Heinrich, refused to leave. "We are getting everything ready," wrote Arnold, "and we very much hope that you will soon be able to hold mommy in your arms."

Emilie feared traveling to a strange land. "If Cuba actually comes about then everything changes, then I'm as good as lost, how sad for me!" she wrote to a daughter, Hilda Steffensmeier, who lived a hundred miles away in Essen and was safe, still, from the Nazis because she had married a Catholic (she would later go into hiding). "Our great worries don't allow for much joyousness," Emilie wrote. "Nevertheless we can hope for ourselves and our family that we reach the island where at least one can live like a human being!" Jews were beginning to starve in Paderborn; friends were being arrested and deported, some resorting to suicide as a last means of escape.

Seeking aid for her relatives trapped in Germany, Wolfgang's mother, Anna, traveled to New York, where a cousin lived—Arthur Nussbaum, the son of Julia and Emilie's sister Bernhardine, who had visited Bad Pyrmont in 1891 with her bothersome husband, Bernhard. Arthur was the famous international law expert, now a professor at Columbia University. He lived on Riverside Drive. "He was an old man," Wolfgang remembered, "very hard of hearing, very, very brilliant, very intelligent eyes." Anna asked him to help her pay for visas for Emilie, Arnold, and Hilde—she needed $1,600 (in 1941 dollars) per person. Nussbaum agreed to give money but said he could provide only enough for Emilie. In Paderborn, the family grew more desperate. "Please, please dear sister take care of us as fast as possible," Arnold wrote. "Keep in mind and do everything for mother, especially whatever you can to allow us to travel with her."

In December 1941, however, that door closed altogether. Nazi transports began to depart from Paderborn "to the East." Emilie and her sons avoided deportation for a time, pleading poor health. In April 1942, they moved into the town's Jewish orphanage—a charity to which Emilie had once been a donor. "I am gradually adjusting to my new surroundings," she wrote to her daughter Hilda in a series of letters reprinted in Naarmann's book.

It's not only the bothersome lack of space; other moments contribute. But when we hear what impossibilities others have to put up with, we just have to be satisfied and compliant.

I really can't complain about the "accommodations" here in the house. Everyone is very nice and pleasant; it is peaceful. Fifteen children have not been brought back from vacation, sad but true! My time is filled with activity. We often have visitors in the evening, but I prefer being alone most of the time because there is much to patch, to sew so I use the evenings for that purpose.

Those small solaces wouldn't hold, of course. Emilie wrote to Hilda again in May 1942.

My dear Hildchen,

These days I have so little good news to report that letter-writing provides no pleasure, neither for the writer nor for the reader. A threatening imminence hangs in the air again. . . . As of June 3rd, the orphanage has to be vacated. . . . The employees have all been dismissed.

What is to become of the rest of us occupants has not been disclosed. . . . In any event we are facing in the near future another change of residence that surely will not turn out well this time. Barracks or something similar! What we all have had to endure!!!

Naturally we aren't very well provisioned; others of our acquaintance are in the same situation; so we really can't complain.

Where are my hairclips that I had earlier in my bag? I look so awful without my clips. . . . I hate to think of having to share a room with several other people.

With love and greetings,
Mother

The move from the orphanage was postponed to mid-June, then July. On July 12, 1942, Emilie wrote her last letter from Paderborn.

My dear Hildchen,

The worry about the near future weighs so heavily on me that I have no peace at all. Now the newest orders have been issued that all elderly people who are not bedridden will have to go to Theresienstadt. . . . It is supposed to be a privileged transport; we will be allowed to take 50 kilograms of luggage, as well as bedding, in addition to hand parcels. Appeals, it is said, are useless. So, now we have to get ready.

On July 28, 1942, Emilie and her two sons, along with the rest of Paderborn's remaining Jews, were taken from the Jewish orphanage to the city's train station. Emilie was allowed to bring food and provisions for two days—anything else would be confiscated—along with her fifty kilograms of luggage. She was allowed to take two blankets, though they would be included in the total weight. Everything else she owned had already been taken: her cash, jewelry, gold, silver, watches (time no longer mattered), personal papers, pension cards, ration cards, lipstick (appearances were of no importance, either), everything except the identity card that exposed her as a Jew. They were given receipts for the things taken from them.

The train carried Emilie and her sons to the nearby city of Bielefeld, where she crammed with six hundred other local Jews into a large lecture hall near the Nazi parade grounds. They slept on benches and chairs in the building. Three days later, on July 31, the train left for Theresienstadt. Emilie's daughter Hilda traveled from Essen to the Bielefeld station to say good-bye. Emilie's sons, Arnold and Heinrich, were devastated, but Emilie climbed in with her "chin up," Hilda told her daughter Helga, who recounted this exchange in a speech many years later. "I have to set a good example for my sons," Emilie told Hilda. They were loaded into the cars, packed body to body. Before the train left, Hilda watched German soldiers haul away the bags that Emilie and her fellow prisoners had packed for the journey. They dumped the contents on the siding, and the train pulled away.

twenty-four

❧ DESTINATION CAMP ❧

*Portrait of an old woman waiting for food, Theresienstadt
Ghetto, by artist and prisoner Leo Haas.*

The Theresienstadt concentration camp was not an official death camp like Auschwitz and Birkenau. Though tens of thousands of people died within its walls, it was a "ghetto" and transit camp—a holding spot to gather and prepare prisoners for transport "to the East." Located in what is now the Czech Republic, about sixty kilometers from Prague, it held Czech Jews and political prisoners first, then

Jews shipped from Germany, Hungary, and Denmark, among other countries.

I hadn't known of that particular camp until I heard from Wolfgang that Emilie had been sent there; now I wanted to learn everything about it. So after Margit showed us the sites of Emilie's final humiliations—the electroplating factory that sits on the land where her grand house once sat, the train station from which she left Paderborn for the last time, the Jewish cemetery that should have held her remains—my mother and I set off for Theresienstadt.

I wasn't quite sure how Emilie's story linked to her sister's, except that they had both died sadly, whatever side of the Atlantic they were on and whatever steps they took to evade their fates. I had read the Holocaust books that left me with a days-long hangover of anger and sorrow: *Night*, *Maus*, Anne Frank's diary. But to have found a sister of Julia's who had her own role in this most haunting of Jewish stories made the consequences of Julia's heritage—and my own—suddenly real to me. Had Julia not come to America, this might have been our family's fate.

My mother and I took a train from Paderborn to Prague, past Lügde and Bad Pyrmont, through Bielefeld and the rolling Weser uplands, then across the flat Northern Lowland and up the craggy Elbe River valley. Emilie would have traveled the same route on her last journey. In Prague, we spent an afternoon jockeying the narrow streets with the tourist hordes—the Charles Bridge, the Gothic excess of Saint Vitus Cathedral, a fifty-koruna spin through the shabby entrance hall of Franz Kafka's first home—and then signed up for a tour to Theresienstadt.

It was drizzling the next morning; the eternal steely mist of central Europe, the sky the same dirty gray as the stone and concrete buildings and streets. There were only four of us waiting for the bus: my mother and I, and a glamorous Canadian couple on their honeymoon. Our guide was named Oleg. The name fitted him. He seemed to inhabit a certain eastern European archetype: pale and rangy in blue jeans and a black faux-leather jacket, his brown hair flecked with gray, his eyes sorrowful and cerulean. He herded us onto the minibus.

We set off with a screech, careening across a bridge, past the fairy-tale spires of the Staré Město and on toward the city's outskirts. Oleg sat in the front seat and gripped a microphone in both hands. "There," said Oleg, "is biggest statue of horseman in world. Sixteen meter tall." He sighed. "On left-hand side, interesting church." He did not look to the left as he made that observation. He held the microphone so close to his mouth that we could hear spit flecks bounce. I wondered if I should have done more research before booking the tour.

We sped onto the highway linking Prague to Theresienstadt. Oleg launched into a brief, barely intelligible history of the camp. Theresienstadt, he said—I paraphrase, heavily—had been built as a fort in the late eighteenth century to protect the road to Dresden. In World War I, it had held political prisoners—Gavrilo Princip, the Serb who shot Austria-Hungary's Archduke Franz Ferdinand and ignited the war, had died there in a small, dark cell. After the Nazis invaded, they walled off the prison and surrounding town, expelled the residents, and turned the enclosed space into a Jewish ghetto. All told, 150,000 prisoners went to Theresienstadt; only 3,600 people survived until liberation. While most went on to die in Auschwitz or other death camps, more than 33,000 people perished in the ghetto itself—because, Oleg explained, "these conditions of the ghetto and political prison were absolutely non-acceptable for the human life."

Oleg fell silent. He seemed existentially weary—or maybe hungover. The drizzle had turned to rain, and the driver cranked the heat and cracked his window to stay awake. We were lulled by the road and the unbroken gray, my mother nodding off in the seat in front of me, the Canadians leaning in against each other. Oleg dropped his microphone into his lap and fell asleep, too, waking as we turned onto a smaller road and mumbling something about collective farms, and perhaps the Prague Spring. "And on the right-hand side," he said, though we couldn't see the right-hand side, because our windows had fogged over, "you can see the very typical small houses, after the armies in 1968, um"—long, woeful pause, twenty, thirty seconds—"about more freedom. But now we are in the Theresienstadt!"

I wiped clear a spot on my window. We were passing a railroad siding. This was where the trains left for Auschwitz and Bergen-Belsen, Oleg told us. We sped past an unremarkable blur of concrete and fog and continued on into the town of Terezín—the former ghetto of Theresienstadt. The streets ran on a martially precise grid: not a curve in sight; street, building, street, building, street; thick walls, big, blocky entrance arches, flat-faced, stark, symmetrical. There was a rectangular park in the middle, smeared with fallen leaves. A woman with a baby carriage walked hunched against the rain. "Now we are transcending the border between the Jewish ghetto and the small fortress," Oleg explained as we sped past the town and pulled into the prison grounds. It was rather a poetic way to put it.

The bus parked next to a vast cemetery, row upon row of low rectangular grave markers presided over by an enormous Star of David at one end and a cross at the other. "Attention of the hat," Oleg said, motioning us down the stairs of the bus. We walked to the entry of the fortress, where we were to join a larger group of tourists with another guide before rejoining Oleg to visit the Ghetto Museum in the town. As we waited for the tour to start, Oleg asked me why I was taking notes. I told him I was writing about a relative who had been deported to Theresienstadt. He looked unhappy. "It's a very sad place, very gloomy here," he said. He pulled his collar up against the rain. "Not so often I go here, but every time I feel it bad."

The tour of the small fortress made me feel it bad, too. It was as horrifying as one would expect—the bone-chilling cold, the dark, the dank, the claustrophobic cells, the torture and execution sites, the German slogan over the entrance to a courtyard: ARBEIT MACHT FREI, "Work sets you free." But it seemed rather sterile and unreal amid the bustle of our group of Holocaust sightseers and English schoolkids. I couldn't decide if the tour was educational or macabre, valuable or mercenary—nor where my own pursuit of Julia's and Emilie's stories lay on that spectrum. I wondered if I should feel the desolation of the place more strongly for knowing that my great-great-great-aunt had died there. And I wondered if revisiting these sad truths restored the humanity of the lost, or if it simply served to gratify the tellers and the listeners; if the act of retelling kept those who suffered imprisoned in their unhappy endings.

We watched snippets of a propaganda film the Nazis had made about Theresienstadt. This was their model camp, their *Musterlager*, to prove to the Red Cross and international delegations that they treated prisoners well. Prominent Jewish artists, musicians, professors, and functionaries were brought to Theresienstadt. The elderly were shipped here, too. The Nazis had claimed they were deporting Jews to camps in the "East" to perform hard labor, but it hardly seemed plausible that people as old as Emilie could work; thus the Nazis presented Theresienstadt as a "Jewish retirement ghetto," a "holiday camp" where, as the Nazi commandant Heinrich Himmler explained, elderly Jews could "receive their pensions and benefits and . . . do as they will with their life . . ." Upon their deportation, those who still had homes had to sign "purchase contracts" exchanging their houses, their remaining money, their life insurance policies, and their belongings to the German state in exchange for room, board, and health care for life.

Theresienstadt appeared, in the film, to be a pleasant place to while away a genocide. It was, the Nazis said, an idyllic lakeside spa "settlement." When foreign delegations insisted on visiting, the Nazis shipped the sick and malnourished to the death camps and beautified the place for a time—planted gardens and flower beds; painted houses; built a playground; opened a café, a music pavilion, and a community center; and created a special Jewish currency (a hook-nosed Moses, holding the tablets of the law) so that prisoners could go "shopping" in stores stocked with items confiscated from arriving prisoners' suitcases. To soften the numerical grimness of the place, they gave the streets names: "L1 Strasse" became "Lake Street," though there was no lake.

We sat in grim silence—the schoolkids, the ashen-faced Canadian honeymooners, my mother, and I—as the film showed smiling Jews, children giggling, eating, women gardening, a soccer match, a chess game, a lecture, a concert. "The organization of the leisure time is left to everyone's discretion," said the narrator. "I'm all right in Theresienstadt," added the voice of a young prisoner. "I don't miss anything."

After the film, Oleg guided us back onto the minibus, dropped us at the Ghetto Museum, and wandered outside to smoke. The museum was situated in what had been a boys' home during the war, number L417. Theresienstadt had had its own Jewish administration, which worked to provide decent conditions for the children, placing them together in dedicated houses and trying to keep up their education and their spirits. Imprisoned artists and academics volunteered to teach the children, believing, still, that learning could salvage hope. The artists and the children sketched their sorrow. The musicians composed symphonies. An amateur chorus performed Verdi's *Requiem* for the prisoners and their Nazi jailers. They had one piano and only a single score, so all the singers worked from memory. They sang their own requiem.

In the downstairs rooms of building L417, the walls were lined with children's drawings that one of the art teachers had preserved before she was deported to Auschwitz. There were dragons and princesses carefully outlined and colored, alongside SS guards and watchtowers—

hope side by side with despair. Some were scrawled and simple, some more sophisticated. They were drawings by the doomed: a sign informed visitors that of the more than ten thousand children who passed through, only a few hundred survived.

Next came the crematorium, an iron contraption in a barnlike building where the bodies of Theresienstadt's dead were reduced to ash. A bas-relief sculpture of an owl regarded us from above the door. "Maybe is symbol of immortal memory," Oleg suggested. We explored the stark rooms, clutching our raincoats closer around us, chilled by the statistics: the average death rate by September 1942 was eighty prisoners a day. Bodies arrived at the crematorium with a tag attached to one leg—name, transport number, group number—and left as dust. The crematorium was capable of handling almost two hundred bodies a day. At first the Nazis stored the ashes of the individual dead in wooden urns, and then in paper cinerary bags; later, they began incinerating the bodies together. By the war's end, they stored the commingled ashes wherever they could, Emilie's among them. In 1944, with the Allies approaching, the Nazis emptied 22,000 urns of those ashes into the river Ohre and buried another 3,000 in a pit nearby—a bulldozer operator found the macabre trove in 1958.

We were running behind schedule; Oleg fairly pushed us onto the bus. We drove through the fallow fields surrounding the camp, a few husks still standing, and back to the pitched roofs and curved roads that you find in real villages where real people live, where things are cluttered and asymmetrical. The Prague skyline grew visible through the drear. Oleg, caressing his microphone, pointed out the horseman again, still "sixteen meter tall," and a very tall television tower. I could see Wenceslas Square approaching, where Oleg would gratefully leave us. He began talking much faster, almost cheerful now, his words skipping away: "On the right-hand side the embassy of Brazil, on the left-hand side the embassy of Argentina, now we will finishing our trip today, I would like to say you welcome again Prague, welcome the Czech Republic, try to be careful, and good lock."

◇◇

Emilie and her sons arrived in Theresienstadt in a wave of "elderly transports" from Germany—during the second half of 1942, 124 transports brought nearly 31,000 German Jews to the camp. The next year, the Germans built a siding directly into the camp. But until then, prisoners had to walk three kilometers from the railway station outside Theresienstadt to the ghetto—old, infirm, no matter. Emilie, age eighty, walked those long kilometers. And when she passed the ramparts, gates, and concertina wire that marked the perimeter of her new "destination camp," she found no hot springs, no water cures, no lake—just people and more people, starving, filthy, doomed. In August 1942, the town's stern grid, designed to house 5,000 people, held 50,000 instead—the next month, it would contain nearly 60,000 prisoners. That summer, 2,000 people were arriving each day. Almost half of the prisoners were over the age of sixty; more than half of those were women.

There was nowhere to put them. They slept wherever they could find a spot—in attics and dark corners, on the ground and in cellars, in sheds and converted stables and pigsties and hallways and entryways. There was no lighting; the electrical system had collapsed. We don't know if Emilie was allowed to stay with her sons or if she was placed in a "home" for the elderly. Either way, she could have expected to occupy about five feet of floor space. If she was lucky enough to find a bed, it would have been a bunk three slots high, each bunk two feet wide, with two and a half feet of head space—too tightly stacked for sitting upright. In the bunk rooms, seventy prisoners might occupy a space designed to house ten soldiers when it was built. The elderly who arrived during the summer of 1942 weren't typically assigned a bunk, however: most slept on the floor or on their own suitcases if they were lucky enough to arrive with one. Memoirs from the camp described prisoners sleeping on rotting straw mattresses vacated by the dead, riddled with maggots and bedbugs, and gnawed by rats and mice. Or on piles of wood shavings. The walls dripped with moisture.

There was a desperate shortage of water, of sinks, of toilets. On July 20, shortly before Emilie's arrival, the sewers stopped working entirely. The prisoners instead used latrines and outdoor ditches, whose smell caused even the hardened to heave with revulsion. Nonetheless, there were always long lines to use them. The prisoners bathed in icy water, if at all. They were allowed to wash three kilograms of laundry every six weeks. Dirt bred infestation—they were covered in bites: bedbugs, fleas, lice. There were typhus outbreaks, dysentery, rooms of people lying in excrement, one body on top of another.

They were starving. The Jewish self-governing council rationed what little food was given the prisoners. The committee steered larger portions toward the young and more able, who worked in the camp's factories eighty to a hundred hours a week, making wood boxes, splitting mica, and sewing uniforms. Those registered to work received fifty grams of sugar, fifty grams of margarine, and a quarter loaf of bread each week. Those who didn't work—Emilie would have been among them—got ten grams of each. They ate rotten potatoes, soup that was nothing but warm water, and on good days a gruel made of watery coffee with pieces of margarine and potato starch. They'd stand in an endless line for a cup of ersatz coffee and a thin slice of low-grade bread, covered in mold—and Emilie had owned a mill! Fights broke out over a few grains of millet. People stole from each other, old and young. The elderly suffered greatly—carefully dressed old ladies with nets in their hair, men in pinstriped trousers wandering the streets and halls begging for bread and soup, holding their receipts, proof of the money they had paid for the privilege of living in a sham retirement camp. In August, Emilie celebrated her eighty-first birthday.

In September alone, nearly four thousand people died in the camp, most of them elderly. They starved or succumbed to disease. To relieve the overcrowding, the Nazis dispatched eight elder-transports to the east. Of the 16,000 who left in the fall of 1942, only one man survived—he leaped from a train near Dresden. But Emilie avoided those transports and survived the difficult summer and fall.

Then winter descended. A very few of the rooms housed stoves fueled with sawdust. The attics and stables, cellars and hallways had no heat. A lucky few still had the blankets they had brought with them; most had only the clothes they had worn on the train. Their overcoats, thin for such cold, grew even thinner. They died from malnutrition; they died from typhus, from typhoid, from tuberculosis, cold, despair—too many homeless ghosts to hold in the mind. They might stay frozen in place for days, the living threading through a maze of dead to leave the room.

The bodies of the dead would be heaped on two-wheeled carts, rolled through the streets, stiff and gray, their legs and arms jutting out. Each day, wagonloads of corpses were wheeled to the crematorium. When Wolfgang first told me about Emilie, he said she had died from "exposure." Exposure to the elements, I supposed; exposure to the worst, too, that humans can inflict. Her death certificate was more particular. The Nazis were precise about the details of death: she perished from "enterocolitis"—dysentery. Her *Sterbetag*—date of death—was January 1, 1943; her hour of death, eight in the morning.

Emilie had lived almost fifty years longer than her sister Julia—who, in all her sadness, had been lucky enough in the end to die with a house to haunt. Emilie had prospered in life, but she had died amid unimaginable horror: the last of the sisters, the last of her generation.

Her ashes are lost, intermingled with those of so many others who perished at Theresienstadt. There's a small plaque with her name in the Jewish cemetery in Paderborn, next to the more substantial grave of her husband, Louis. TO DEPART IS THE FATE OF ALL PEOPLE, reads a tombstone not far from the family plot. We are all mortal, of course. But how we depart: it makes a difference to those who remain.

Julia could not have imagined Emilie's fate, even in her worst moments. Nor could Abraham, or Bertha, though it could so easily have been them or their descendants, had they remained in Germany. In all the pages of Bertha's diary from 1891 and 1892, through all the spas, dances, family visits, sad moments with her mother, and encounters with "young gentlemen," Bertha doesn't mention once, not once, ever, the fact that she was Jewish. There is no discussion of synagogues, or Sabbaths, or anything that would suggest religious observance. The only holiday she mentions is Christmas. Being Jewish wasn't something she seemed to dwell on. And yet fifty years later, Bertha's aunt and cousins in Germany were nothing, nothing at all, but Jews. For that, they lost everything.

There's a bicycle trail that now connects Paderborn to Neuhaus, through the meadows Emilie once owned. It is named after her: "Emilie Rosenthal Way," a manicured path of atonement that wanders along the river Alme. The river grasses shimmer green, and the water flashes jeweled glints of sun between the shadows of overhanging plane trees. On the spot where Paderborn's Byzantine-revival synagogue stood before it burned on Kristallnacht, a plaque lists Emilie's name and those of more than a hundred other lost Paderborn Jews. Below the names is a lamentation. It comes from an Old Testament dirge to the dead of fallen Jerusalem: "Is it nothing to you, all ye that pass by? Behold, and see if there be any sorrow like unto my sorrow, which is done unto me."

Two months after Emilie's death, Heinrich was able to send a postcard to his sister relaying the sad news.

Dear Hilda!

Unfortunately I must send you the inexpressibly sad information that our dearly beloved mother died on January 1 after a short illness, after charging me with the warmest farewell greetings for you, which I hereby discharge. The memorial service took place on January 4. I miss the dear departed infinitely. My pain is like yours. I spent my birthday just as sadly as you will experience yours. I wish you, however, the best for your new year of life. I've received the contents of your little package. Warmest thanks.

With sad greetings I remain your faithful brother,
Heinrich

Theresienstadt, March 4, 1943

Heinrich was sent to Auschwitz on March 29, 1944. He died there, as did Arnold and Hilde—betrayed by Germany, failed by history. Of the one hundred and twenty-three Jews who were still in Paderborn in 1939, only five survived the Nazis. None survived in Lügde: its last Jews were sent in 1941 to Riga, Latvia, where they were shot.

They haunt us still.

Juli

◇◇

I MET JULI IN the library of the old house at La Posada. She was tall, thin, straight-haired, and blond, with small gold teardrop earrings—sensible looking. Juli ran a psychic institute in Santa Fe, and I'd asked her to meet me at Julia's house.

The library was stunning, filled with avant-garde Southwestern art, the floors a delicate inlaid parquet. We sat on a cushy couch near the fireplace, where the family had gathered on chilly evenings more than a hundred years before, and Juli told me what she saw: a woman with long, wavy, white hair, wearing an old-fashioned cotton nightgown. The woman wandered the upper floor: "The sense I get is that she's looking for something, and the other sense I get is that she's supposed to be entertaining people. It feels like it's an anxiety for her, a pressure, and she doesn't have it within her to be doing that," Juli said.

As I looked around Abraham's elegant library, this made sense. The house itself—the grandness of its construction—was surely a player in Julia's suffering. It was full, then as now, with people expecting Santa Fe's finest treatment—Abraham's railroad and poker friends then; gallery-goers and high-flying tourists waiting for spa treatments now. The house carried too many expectations.

There was a sense of peace about Juli—a competence and restraint that made me want to believe her. She saw an image of Julia getting ready for an event, putting on her evening clothes, brushing her hair, getting ready, getting ready, but unable to go down the stairs. "I keep getting a sense of this overwhelming pressure that she felt," Juli said.

Juli sat calmly with her hands on her thighs, her eyes closed. "What I get is that she didn't have a real strong constitution to begin with," she said, and all the children, all the miscarriages, sapped her vitality even more. "I see her getting physically weaker with her births, and yet there were still all of these expectations on her."

Julia's body grew weaker; her psyche grew weaker. The rope that connected her to the world grew more and more frayed. By the time Julia lost the baby, she was already "off," Juli felt— "she wasn't able to care much for another being." Abraham couldn't handle Julia's sadness—he didn't understand it. As she fell apart, he grew more absent. Julia felt abandoned. "The image I get is that she was really lonely here. I see this deep loneliness," Juli said.

The stories I was hearing, again and again, the speculations on Julia's life and death—they did seem to cohere. There were elements that repeated themselves: Abraham was a solicitous husband but a difficult one, with his high expectations and extravagant standards. The loss of the baby was horrible, but only one of many blows that struck Julia down; she loved her children dearly but found herself unable to care for them. You can be in despair but not be insane. Perhaps these details came together this way because I had shaped them in the asking and the telling. Perhaps it was easy to make these assumptions about Julia's life, because so many nineteenth-century women lost children and suffered from high-handed husbands. Or perhaps they were true.

I asked Juli about the dreadful accident in Germany. Julia wouldn't tell her, any more than Bertha would explain it in her diary. "She becomes very agitated," Juli said. "She just becomes dark. My sense is that it's something to do with her face just by how I see her looking right now. Something really big happened." Now Julia deflected Juli—she didn't want to discuss it. The fireplace hissed quietly. "I'm seeing her now, up in her bedroom," Juli said, "brushing her hair. She's being pressured to entertain guests, and she breaks. She comes down the stairs in her nightgown, looking very ragged." I could see it, too. Julia, a ghost of herself already, half in the world and half out, swaying unsteadily on the stairs that ran beside the room in which we sat.

"She comes down and thinks she's hosting all these people," Juli said. "It's funny, it's that connection with this being a public place that keeps her here, in a sense—that this is her job to be the hostess. I keep seeing her walk down a staircase—she keeps thinking she's showing up, dressed beautifully. She doesn't realize she's dead."

twenty-five

❧ HER LONG REST ❧

Julia.

After her accident, Julia remained in Hanover. In January 1893, almost a year later, she was, reported the *New Mexican*, still unwell: "In case Mrs. A. Staab is able to travel she and her daughters, now in Europe, will leave there about April next and summer in Santa Fe." She didn't leave April next, however. In May, the paper found Julia still in Hanover, "and the latest news is that she is steadily, if slowly, improving." In July, Abraham went to fetch her. They departed from Bremen in mid-September on the steamer *Havel.* Abraham and Julia traveled directly from New York, arriving in Santa Fe on October 6. Bertha detoured to visit the Chicago World's Fair on her way home—the fair at which the historian Frederick Jackson Turner would take the podium to declare the West officially settled, the frontier formally closed.

Julia's own borders closed in on her that year as well. She disappeared from the public eye. I suspect this was a relief for her, not having to try anymore. She stayed in her room now, locked away, the curtains on her four arched windows drawn against the sunlight and the world. Not once, after Julia returned, did "Mr. and Mrs. A. Staab" appear on the society pages. Now it was "Mr. A. Staab and Ms. Staab"—Bertha, who assumed the role of Abraham's official hostess.

Bertha was the last of the maiden Staabs left; Delia had become engaged to an Albuquerque wool merchant—another German Jew— soon after they returned from Germany. They married in November 1894. "The Interesting and Beautiful Ceremony Which Last Night United in Marriage Mr. Louis Baer and Miss Delia Staab," read the headline in the *New Mexican*, which reported it as "one of the most brilliant social events that ever occurred in this city." The territory's chief justice presided over the ceremony, held, as Anna's 1889 wedding had been, in "the handsome Staab residence of Palace Avenue," decorated with heliotropes, chrysanthemums, white roses—"flowers and vines throughout." Delia "a vision of loveliness," came into the parlor "leaning upon the arm of her father, Mr. A. Staab." The groom, however, was "attended by Mrs. B. P. Schuster, of Albuquerque"—a cousin-in-law—"who took the place of the bride's mother in her absence on account of illness." Julia stayed in her room. I imagine her as Juli described her: at the top of the stairs, half-dressed and unkempt, close to family below but no longer in their world.

Abraham, of course, hadn't retreated from the world one bit. In the

months after Julia's return, he fought to keep the territorial capital in
Santa Fe, fending off an effort to move the territory's political center of
gravity to Albuquerque, which had by then become the region's economic center. The battle was ugly; Abraham's character and economic
motives were frequently maligned. In February 1895 he holed up in
the capitol with Catron and a group of like-minded legislators flanked
by fifty armed guards; Abraham personally guaranteed the funding to
rebuild the capitol building and a nearby penitentiary, and to arrange
a second railroad spur directly to the capitol from the train station in
Lamy, outside town. When the vote fell in Santa Fe's favor, Abraham
was exultant. "For once," reported the *New Mexican*, "Mr. Staab lost
all his dignity today . . ." In his honor, the city named the narrow road
to the railroad depot "Staab Street." It is short—only two blocks—but
it lies at the city's heart. Catron Street, which runs parallel to it, is much
longer. But I get the sense that Abraham never minded public eclipse,
so long as his private concerns were satisfied. He continued to serve on
the Santa Fe Board of Trade, representing the city's interests in Washington, New York, and elsewhere.

So it happened that he was gone, representing the Board of Trade in
New York, on the night Julia died. It was May 14, 1896, late in the evening. Springtime: the desert was etched in pale yellows and pinks, like
a woodcut. "Death of Mrs. A. Staab," read the *New Mexican* headline
the next day. "The Wife of the Well Known Santa Fe Merchant Called
to Her Long Rest Last Night." She "died quite suddenly at 10:30 last
night," the newspaper said, "after a protracted illness extending over a
period of five years." Anna, the oldest daughter, took over the prayer
book, inscribing the details of Julia's passing. It was a Thursday: "*Don-
nerstag*," she wrote. I wonder if Julia's death wasn't, by the end, a relief.

I could find no explanation of the particulars of Julia's death. The
newspapers offered no specific details. Nor were there official death
certificates at the time. Dr. Harroun's notebooks offered no infor-

mation; Bertha's diary had petered out three years before. Grasping
for any insight—any imaginative flesh to plump up the unsatisfying
factual bones of the story of Julia's last days—I'd asked the psychics
for details. Ed Conklin had said heart disease was involved. Lynne
believed it had happened in the bathtub. Juli had thought it was an
overdose of medicine, an unintentional poisoning. Misha, the phone
tarot-card reader, didn't give a specific cause, but assured me that it
wasn't suicide. Sarina had wondered whether Julia might have suffered
from a sexually transmitted disease, but she believed that the immediate cause of death was associated with a lack of air. "She went like
this," Sarina had told me, grabbing her neck, as if gasping.

I imagined Julia, her white hair unleashed and wild, wraithlike
in her nightdress, downing too much laudanum and climbing into
the bath, her heart clogged and slowing, sinking into the tub—her
beloved porcelain bathtub, toted across the prairie—sinking below,
gasping for breath, water filling her lungs, fear and relief and regret
mingling in those last moments as she sank below, too enervated to rise
to the surface.

Whatever happened to her in the end, the death was news in Santa
Fe. A lengthy story in the *New Mexican* recounted what we know: born
in Lügde, one of twelve; maiden name, Schuster; seven children left
behind; gone too young at fifty-two. "She was a woman of many noble
characteristics, a devoted wife and a true mother. Telegrams were last
night sent far and wide announcing the sad news." The first of those
telegrams went to Abraham in New York, "informed by wire of the
sudden and unexpected death."

We can lay one rumor, then, to rest: Julia did not die by Abraham's
hand, as some versions of her story have suggested. A sudden death,
after a long decline, and Abraham more than half a continent away.

Despite the Jewish requirement that bodies be buried no more than
a day after death, the family delayed the funeral for five days so that

Abraham and the youngest boys, Teddy and Julius, both in school at Harvard, could come back to attend the services. They took the first train home. "The body has been embalmed by Undertaker Gable and is at the family homestead under the watchful care of the sorrowing relatives and friends. A fine casket has been ordered from the east and will arrive here to-morrow night," said the *New Mexican*.

Stores closed throughout town for the funeral, which, like many of the important events in Julia's life, took place at the house on Palace Avenue. "The ceremonies," reported the newspaper, "were of the most impressive character." A family friend read the Jewish burial service "with grave and stately mien." A hymn was "sweetly sung" by four Irishwomen. "Rare floral tributes" covered the casket, which was removed to a horse-drawn hearse and then to Fairview Cemetery—the new, nondenominational graveyard on Santa Fe's outskirts. Abraham had helped fund it so that there would be a suitable place in the city for non-Catholics to be buried. Here Julia was lowered into the red, gritty, still-unfamiliar New Mexico dust, and Kaddish was recited, the men facing east, to the mountains, to the plains beyond, to the sea, to Germany and Jerusalem. Julia was consigned, once and at last, to the earth, and to the past.

And then life continued, as it does. Teddy and Julius returned to Harvard to finish their exams—Teddy graduated in June. Anna and Delia returned to their husbands and families in Albuquerque. Abraham presented Julia's last will and testament; the newspapers mentioned no surprises. Six weeks after the service, he and Bertha—his last unmarried daughter—left for Germany. They spent the summer in Carlsbad, returning only in October. On his return, Abraham visited his married daughters in Albuquerque. He sued a few people—mourning did not, apparently, restrain his litigious disposition. He was likewise sued: by customers, landowners, poker buddies. He traveled to Washington ("Hon. A. Staab, the merchant prince of

Santa Fe . . . boarded the train . . . for the Capital City"), hoping to convince Congress to pass a bill that would fund his outstanding militia warrants—he hadn't given up on the militia warrants; he wouldn't do that. He went about the business of being a merchant prince.

In 1901, Abraham was included in the *New York Herald*'s annual list of millionaires—one of only four in New Mexico. In 1903 he appeared in the *Herald* again—this time, the newspaper listed his occupation as "retired." He had sold the dry goods business to two local merchants. With the army gone from Fort Marcy and the railroad bypassing Santa Fe, the territory's large wholesale companies were all now based in Albuquerque; its population now surpassed Santa Fe's for the first time.

Santa Fe was losing people, in fact—its population dropped by more than a thousand between 1880 and 1910. Abraham's had been the last big mercantile house in the city, but he was nothing if not a savvy capitalist, and he understood that the dry goods trade was a dying business model—as the frontier had receded, so had the need for his goods. So he had diversified over the years: bought hotels, saloons; expanded his real estate interests; invested in railroads, gasworks, mines, cattle, irrigation projects. When he "retired," he held on to his newer interests.

It was Abraham who now pulled the family together—and I suspect that this had probably been the case for some time. His children had always revolved around him, like planets. When I'd interviewed Betty Mae, she had told me that it was Abraham who had decided all of his sons' careers; this wasn't at all unusual in that era. Paul, the oldest son, was in no condition to work, so Abraham picked Arthur to help run the business. Julius became a lawyer—he practiced in Albuquerque, overseeing many of his father's legal concerns; Teddy's designated profession was medicine—he stayed east as a pediatrician in Philadelphia.

After Julia died, only two children remained home with Abraham: Paul and Bertha, who served as her father's hostess and accountant—

part dutiful daughter, part serf. "Grandfather ruled over her with an iron fist," Aunt Lizzie wrote of her mother. Whatever inadequacies Bertha professed to her diary, she had never suffered from a lack of suitors: she was engaged three times, Lizzie said, but Abraham broke up each affair. If Bertha ever reunited with the man she pined for while traveling in California, she left no record of it. Lizzie reported that at one point Abraham tried to force Bertha to marry a man whom everyone in the family "disliked thoroughly." Bertha disliked him, too; she fled to a friend's house to hide out for three days when he came to town. I found a newspaper announcement of Bertha's engagement, in 1902, to a Peoria, Illinois, man named Adolph Woolner; I suspect it was Woolner whom Abraham had foisted upon her. Finally, a brother-in-law intervened with Abraham to break off the engagement.

Bertha was thirty-seven when, in 1907, she married my great-grandfather Max Nordhaus—a German Jew from Paderborn who managed the dry goods business of his aging brother-in-law Charles Ilfeld in Las Vegas, New Mexico, and on whom Abraham and Bertha could finally agree. The wedding, held at the house, was a much smaller affair than either of her sisters'. "Elaborate and elegant, though quiet was the Nordhaus-Staab wedding Thursday evening," reported the ever-faithful *New Mexican*. It was a civil ceremony, attended only by relatives and "very intimate friends."

It would be the last wedding held in the mansion on Palace Avenue, the last big family celebration.

twenty-six

❧ BEQUEST ❧

Abraham died a very wealthy man.

Abraham lived another seventeen years after Julia's death. He remained vital into his seventies, traveling often to California, New York, and Germany. He never remarried. As his children drifted into adulthood, he was a "lonely man," according to Amalia Sena Sánchez, who in an oral history remembered meeting him on a train platform when she was a girl at the turn of the twentieth century. "He used to go to the station just to talk to the people going through."

In 1907, just after Bertha's wedding, Abraham was injured in a train wreck on his way to Denver. He was, the newspapers reported, washing his hands in the "lavatory" of Santa Fe Train Number 1 when the second section smashed into the first and his car crumpled around him. He suffered abrasions on the head and face and had to be cut out of the train. He recovered well, and continued to make his annual summer trip to Europe, stopping at the Brown Palace Hotel in Denver after his visit in 1910. "I've been pretty well over the continent this time," he told the *Denver Post*, "and I have found no city that compares with Denver in point of cleanliness and evidences of prosperity and thrift." Still a charmer, he was.

In late 1911, he bought the territory's first "maharajah-suited automobile," according to the *New Mexican*, a Pierce Arrow limousine— the model favored by Hollywood stars and royalty. Abraham had the vehicle shipped to him on the Atchison, Topeka and Santa Fe Railway along with a chauffeur and trained mechanic from the company's headquarters in Buffalo. The car was red, twelve feet long, with glass windows, velvet curtains, and leather cushions—an exact reproduction of President Taft's own limousine. "It is a beauty of the touring landau type, has six cylinders of 48 horse power, and cost $6,800," the *New Mexican* effused. It was the largest car that had ever been seen in New Mexico. Abraham had a two-story garage built to hold it. He would, at first, allow it to be driven only between his house and office—a distance of three or four blocks—and only in second gear.

The next year, New Mexico won statehood at long last. It had taken sixty-two years. While other territories had entered the union within months, New Mexico and Arizona, with their large Spanish-speaking populations, were deemed unready, the territory's bid for autonomy bogged down in the politics of race and the battle between the Santa Fe Ring—which supported statehood—and its political opponents. But in January 1912, after years of false starts, New Mexico became the union's forty-seventh state. The territory that Abraham had helped build—a once foreign land of Indians and Spanish and Mexicans, and later Anglo-Americans and German Jews—was finally and formally a part of the United States, its years as a territory paralleling Abraham's own American journey.

Abraham was surely jubilant, but his celebrations were tempered now by discomfort. His health had begun to fail. He suffered from heart problems and from "uraemic trouble." Teddy, working as a doctor in Philadelphia, closed his practice and moved back to Santa Fe to treat his father. When Abraham recovered, Teddy moved to New York to start a new practice, but he was called back to Santa Fe when Abraham again grew ill. In late 1912, Teddy and Abraham traveled to Pasadena, California. "I wish to escape the very cold weather at my time of life," Abraham told his friends. "I guess a man can take a vacation in California." This vacation, however, also included bladder surgery, from which he was expected to make a full recovery.

He didn't. On January 4, 1913, three days after his surgery, Abraham suffered a heart attack and died. "Abraham Staab Is Claimed by Death in Hospital at Pasadena," the *Albuquerque Journal* screamed on its front page, in inch-high letters. The news of his death spread across the Western newspapers—San Diego, Denver, Colorado Springs, Las Cruces. Abraham's fortune was thought to be the largest in New Mexico, the papers noted, at more than a million dollars. "Richest Man Dies," read the headline in the *Anaconda Standard*.

Teddy accompanied the body back home, passing through Albuquerque on Santa Fe Train Number 10. "Innumerable family connections are plunged into mourning," wrote an Albuquerque society columnist, "and many social affairs are, as a result, indefinitely postponed." The Ilfeld family "bal masqué" was crossed off the social calendar, as was another society dance. Abraham had, of course, left detailed directions for his funeral. "It was his wish," said the *Journal*,

"that the funeral be held at Santa Fe and that the Jewish rites be observed." A rabbi from Albuquerque's Temple Albert presided. Abraham's body was buried beside Julia's, in the plot at Fairview Cemetery.

Two weeks later, the will was submitted to probate—and this time,
there were surprises. Abraham left the bulk of his fortune to his children, with some "special bequests" and exceptions: $500 to the Sisters
of Charity in Santa Fe, 10,000 German marks to Julia's sister Sofie,
who had cared for Julia during her difficult period in the late 1870s, and
$5,000 to Abraham and Julia's son Arthur. The rest of the estate went
to their other six children; Arthur would receive nothing except the
$5,000 stipulated—"and no more," Abraham declared. Were Arthur
to contest the will, he would not receive even that small bequest. Abraham had disinherited his son. To make the point clear, Abraham made
it again: "I herewith again affirm that my son Arthur shall in no way
participate in my estate under this will except to the extent of $5,000 as
bequeathed to him."

Perhaps Abraham wasn't as despotic a husband and father as the
ghost stories suggest—in Bertha's diary, anyway, I observed genu-
ine affection. But it was clear, from the will and from Aunt Lizzie's
observations, that there was also a fair amount of tyranny. Abraham
brooked no rebellion. There had been major damage in this family:
invalid mother, controlling father, prodigal son. The parents were both
dead now, gone—but their memories still had the capacity to haunt.

◇◇

Arthur had always been the family's "black sheep," according to Betty
Mae. A bon vivant, fond of lawn tennis and football, parties and card
games, he seemed to find duty and decorum more problematic than
did his siblings. Family lore had it that Arthur had seduced the family's red-haired Irish maid as a teenager, resulting in her banishment
from the house. In 1893, he tangled in a bar fight that made the news.

"Joseph Josephs, the saloon man, became involved in a dispute with young Arthur Staab yesterday and the latter received a slap in the face from Josephs. In the police court this morning Josephs was fined $50 and costs," reported the *New Mexican*.

Abraham was fond of parties and card games himself, of course, and he didn't shy from a fight; he forgave all this. What he couldn't forgive, it seemed, was the offense that Arthur committed in January 1904, when he traveled to San Luis Potosi, Mexico, to marry a young woman from Georgia named Julia Nicholson. Arthur had met his Julia when she moved to Santa Fe to live with a sister after the death of her mother. She was Christian, but Arthur fell in love with her anyway. Her family didn't seem bothered that she had married a German Jew; the Georgia newspapers carried news of the engagement and wedding. But the Santa Fe papers didn't—I suspect Abraham saw to that. Perhaps the family wasn't observant, but the girls all nonetheless married other German Jews. The day after the wedding, the *New Mexican* had only this to say: "Arthur Staab is travelling on a sight-seeing trip through Mexico."

He returned to Santa Fe two months later, according to the *New Mexican*, staying with his wife not at the Staab mansion but "at the sanitarium." Soon after, the couple moved to Oklahoma City, where a 1905 city directory finds them working at and living above the "Up-To-Date Steam Laundry." They moved frequently while in Oklahoma—five times, at least, in the years between Arthur's marriage and Abraham's death—living now in rented rooms rather than three-story mansions.

When Abraham died, Arthur learned the news by telegram and rushed back to Santa Fe, arriving five minutes before the funeral services began. The next day, the family gathered for the reading of Abraham's will. It was then that Arthur had learned he would inherit only the $5,000. He quickly contracted with an Albuquerque attorney to explore whether there was any loophole through which he could contest the will.

But before he could take action, he received another blow. He learned that his younger brother Julius had died. In May 1913, five months after Abraham's death, Julius and Teddy had set sail for Europe, traveling to London, Berlin, and Freiburg. Julius then proceeded on his own to a sanitarium in Kreuzlingen, Switzerland, to recover from "stomach problems." A week after checking in, he was found dead in his room.

The doctors contacted Teddy, who wired their sisters in Albuquerque, informing them that Julius had suffered a cerebral hemorrhage. In fact, his death had been a more violent affair, though it would take time for the details to leak out. Julius's death was not what rocked Arthur, however. It was what happened after.

When Julius's will was read, Arthur learned that he had been left out again. Half of the estate went to Teddy, the other half to the sisters. Arthur was astonished and angered. He couldn't understand how his brother could do such a thing. And he believed that Julia's long shadow was to blame.

◇◇

Julius had been the achiever of the family—Julia's namesake, Abraham's pride: commencement speaker at Swarthmore, then on to Harvard and Columbia Law School, a scholar and an athlete. He was a gymnast—excelling in the parallel bars and vault—and an accomplished heel-to-toe racewalker. "Although very short, and thus handicapped, he has made for himself an excellent record on the track," reported the *Boston Herald* in 1894. He was also a coxswain on the college's rowing team, and he excelled at that, too, according to his obituary in the *Harvard Report*.

After Columbia, Julius practiced law in Chicago for two years, then moved to Albuquerque to set up his own practice. He lived at the Commercial Club, an ornate brownstone in the center of town that contained offices, card parlors, a library, a bar, a ballroom, and on the third floor, bachelor suites where many of the city's unmarried comers resided. Julius was a fixture on the social circuit: a voracious bridge competitor, dance partner, and tennis player. He helped found the Southwestern Tennis Association and brought the first marathon to New Mexico's 1910 Territorial Fair. He performed in local shows. "Little Julius Staab was probably the favorite among the blackface artists," the *Albuquerque Journal* said of an Elks minstrel show in which Julius performed. He was also a member of the city's tongue-in-cheek "Men's Fashion League." "Among the instructive papers read at its gatherings was one on 'How to Acquire the Pompadour,' by 'Doc' Moran," said the *Journal*; "another, 'Should Short Men Wear Checked Suits,' [was] read by Julius Staab . . ."

In his late thirties Julius still hadn't married, though he gave the impression that he would like to. The paper reported on a 1911 wedding at which "Mr. Staab was mostly conspicuous for a huge white flower which adorned his buttonhole and was observed gazing wildly about for the prettiest girls . . ." For Christmas in 1912—just a few weeks before Abraham's death, a few months before his own—he told the society pages that he wanted Santa Claus to bring him "a girl."

He was active in local law and political circles as well, serving on the territory's board of bar examiners and attending the territory's Republican convention. He had, reported the *Albuquerque Journal*, "the real gift of oratory." After statehood, he was elected Bernalillo County's first probate judge. But he lasted only a year in the position. In the spring of 1913 he took a leave from his judgeship to travel to Europe with Teddy; in July he mailed his resignation to Bertha, who delivered it to the county clerk. "He gave no reason for his resignation," wrote the paper. He simply said that he would be unable to return to Albuquerque in the immediate future.

A month later, on August 27, he was found dead at the age of thirty-nine. "It is believed that the grief over his father's death and close application to his duties as probate judge and his law practice, weakened Judge Staab's constitution," wrote the *Albuquerque Journal*. The body was cremated in Europe, the ashes shipped back to New Mexico. In late September, the rabbi who had conducted Abraham's service did the same for Julius's.

After Arthur learned that he had once again been disinherited, he filed a formal objection to the will, which Julius had written only a couple of months before he left for Europe. If it were overturned, Arthur would stand to inherit a one-seventh portion of the estate as a surviving sibling. All he had to do was convince a judge that Julius hadn't been of sound mind when he wrote it—that Julius, like his mother before him, had been insane.

In June 1914, Arthur's lawyer took testimony from the doctor who had treated Julius in Switzerland, a Dr. Ludwig Bingswanger, the manager of the sanitarium in which Julius had died. Dr. Bingswanger confirmed that Julius, when he died, had been under treatment for problems of mental health at an "institution for the treatment of nervous diseases." Julius wasn't being treated for stomach problems. Nor, Dr. Bingswanger testified, had Julius died of a hemorrhage in the brain.

No, he had been found in his room, alone, with a gunshot wound to his heart. Under his body, the nurses had found a "small-caliber revolver"—Julius's own gun. It was then, after Dr. Bingswanger's testimony was released to the newspapers, that the rest of New Mexico learned what Julius's siblings by then surely already knew: Julius had shot himself.

twenty-seven

❧ DIASPORA ❧

Uncle Teddy.

The case went to trial in October 1914, and each day's testimony was front-page news, offering the state's newspaper readers—and me, tracing Julia's tribulations through the next generation—a glimpse into the lives of a once charmed family. "Julius Staab a Good Player but Hard Loser in Bridge Game," read the *Albuquerque Journal*'s headline after the first day of the trial. Friends of Julius's admitted, under questioning on the first day, that Julius often played bridge late into the night, and that he was "of an excitable and nervous temperament"— the question being whether he was so excitable as to be insane. "There were certain humorous tilts between counsel and the witnesses as to just how hard a loser a bridge player must be before being considered erratic or abnormal," explained the paper. Arthur's lawyer asked one witness if it were not true that Julius, while playing bridge, would "deliver a lecture almost every time a card was played."

"He would when he was losing," responded the witness.

Julius's mind was, friends and colleagues testified, "sound, clear and vigorous"—but he was also very nervous. He at times "approached a condition of hypochondria," and often complained of stomach troubles: "frequently after eating a hearty meal would show in his face the visible evidence of suffering."

The second day's testimony revolved around the goings-on at the Commercial Club, where Julius lived: "Dead Man's Roommate Interesting Witness," read the headline. That was Ernest Landolfi, Julius's roommate for more than a year. Landolfi didn't think that Julius was insane but admitted that he was "somewhat peculiar." Arthur's attorney asked Landolfi whether Julius "took baths with undue frequency." Landolfi said that Julius always took a bath after a game of tennis, "but would not say whether he took a bath five or six times a day, as suggested by the attorney." The attorney asked if Landolfi had ever told employees at the club that Julius "had wheels" or was "bug-house:" "Mr. Landolfi with an emphasis that amounted to vehemence denied that he had ever made such a comment."

And on it went, Arthur's lawyer attempting to prove that Julius had lost his mind; the family's attorney asserting that he was just a little nervous. The book Julius had used in his work revising the state legal code was "literally covered with memoranda and citations of authorities"— evidence, Arthur's attorney said, of deep mental disorder. McCline,

Abraham's butler, testified that Julius once refused to come to dinner when called, and that Julius once mistook him "for another negro, named Bramlett. . . . 'I knowed something wuz wrong,' " McCline testified, " ' 'cause I'm a lot handsomer man than Mr. Bramlett.' "

Charles C. Catron, the son of Thomas, the Santa Fe Ring leader and US senator, also shared a long, convoluted tale about a 1911 trip to Roswell with Julius, who talked of nothing but his stomach trouble the entire trip. In the car on the way home, they'd broken out a flask of whiskey, then run into a snowbank. Catron had gotten out of the car to push them out. His feet had grown cold, and when he'd climbed back in the car and requested "a little solace in the way of a snifter," he'd "found to his disgust that they had drank up all the medicine." The *Albuquerque Journal* reported a "distinct trace of emotion in Mr. Catron's voice as he related this occurrence." It was not clear what bearing Catron's story of a drunken car wreck might have had on Julius's sanity, much less what happened after the wreck, when Catron decided to test Julius by ordering him a large steak after they returned to Roswell to repair the car. "Judge Staab, [Catron] said, continued to complain of his stomach, but ate the steak, nevertheless." Proof of madness, perhaps—or maybe an unhealthy fondness for whiskey and steak.

Still there was, Arthur's lawyer suggested, a general "taint of insanity" in the family that began with the parents and worked its way down into the children. "Staab Family Eccentricities Aired in Court," read an October 23 headline. "A recital of the peculiarities and eccentricities of various members of the family occupied a good part of the time of the court yesterday." The Staabs' longtime family physician, W. S. Harroun, took the stand: the mental condition of Paul Staab, the oldest brother, was compromised by epilepsy, he said, brought on by an early bout of meningitis. He explained that Abraham, at the time of his death, was suffering from "paresis." I looked this up: "partial inability to move," according to my *Webster's* dictionary; "a problem with mental function due to damage to the brain from untreated syphilis," according to an online medical dictionary. "Paresis," uraemic trouble, miscarriages, meningitis, insanity—all these were consequences of syphilis, which wasn't, until 1909, treatable with anything but doses of mercury—the side effects of which were often as bad as the disease itself. Perhaps Julia had suffered from syphilis, too.

Harroun testified last about Julia's condition. She was, for some years prior to her death, "insane," the *Journal* reported—our first firm diagnosis of Julia's malady. Sadly, the paper—exercising perhaps the same Victorian restraint that Bertha had shown in her diary—omitted further discussion, saying only that Julia's mental illness was "due to causes that he"—the doctor—"described." Hoping to read in the trial records what exactly he described, I called and wrote and emailed to courthouse after archive after courthouse seeking the transcripts, only to conclude, after months of searching, that the records from the trial had been irrevocably lost, along with my opportunity to know the "causes" of Julia's insanity. But here, at any rate, is what Arthur implied: that Julia's sadness was a condition, a trait handed down in the family, like small stature or long ring fingers—her bequest. And that Julius, like his mother, battled demons.

◇◇

The trial was, at the time, the longest civil case in Bernalillo County history. In the closing days, the arguments revolved around whether Arthur's family had turned away from him because they were crazy or because he was dishonest. When Teddy testified, he denied that the family had been worried about Julius's mental state, and described "with infinite pathos," Julius's final trip abroad, how they "had gone from one to another European resort in a vain endeavor to regain his health"—not all that different from Julia's odyssey two decades earlier. "Throughout Dr. Staab's narrative members of the jury leaned forward in their seats to hear what he had to say," reported the *Journal*, "while a hush pervaded the courtroom as the spectators breathlessly followed his story." A family's ghost and demons—its once hidden past—was now on display.

Teddy explained that Abraham had not disinherited Arthur because of his marriage to a Gentile—though Abraham was angry that Arthur hadn't asked his permission, and Teddy "admitted having heard him say that marriages between people of the same religion were always the happiest." Rather, Abraham cut Arthur out of the family, Teddy said, because Arthur had stolen from them. "My father trusted him implicitly in money matters and this trust was often abused. He took money from my father. He stole jewelry from my mother," Teddy said.

Arthur, during his turn on the stand, insisted that any money he had taken had been promised to him to compensate for Julius's and Teddy's expensive Ivy League education—after only two years at Swarthmore College in Pennsylvania, Arthur had been called home to work for Abraham. Abraham had given Arthur authority, he said, to spend up to $10,000 a year. But after Arthur's marriage, Abraham and Julius had turned on him. They had invited Arthur to meet at the house of a family friend, searched him for a weapon, and accused him of stealing money and pawning Julia's jewelry in El Paso. Arthur had signed a confession admitting the thefts, but he explained that he had signed it without reading it—only, he explained, because his wife was sick and he "did not want to cause her a shock."

Abraham hated Arthur without reason, Arthur said. This German Jewish patriarch could not abide his too-American child, and he dangled his inheritance as the price of compliance. "The witness told a story . . . of hostility to him and favoritism to his brothers from his early youth—of fidelity to the interest of his father which was repaid by unkindness and bitterness." Abraham had once threatened to disinherit Julius, too, until Arthur "pleaded long and earnestly with his father to 'give Julius another chance.' " Abraham's displeasure had a fearsome price.

This was why, when Arthur decided to marry, he told Abraham only that he was "going south." He knew his father would oppose marriage to a Gentile, Arthur said, because Abraham had "broken up a former love affair . . . on that account." After the marriage, Arthur didn't see his father again until a month after he and his new wife returned to Santa Fe. He was sitting with her on a bench on the Plaza as Abraham walked past. Abraham at first pretended he didn't see them. "Hello, Papa," Arthur called. But Abraham said he wanted "nothing to do with him," and walked on.

Like Ishmael in the desert, Arthur was banished. He and Julia moved to Oklahoma City and opened a laundry. "Mrs. Staab, slender, graceful, and possessed of the soft voice that is typical of the high-bred southern woman"—testified about their tribulations in Oklahoma. Arthur's savings had been lost in a Texas bank failure (though in his own testimony, he hadn't been able to remember the name of the bank or the town in which it was located). At the laundry, they worked long hours, on Sundays, and at night, "and on some occasions until 5 o'clock in the morning to keep things going." Julia had to pawn her rings to raise enough money to make payroll. She miscarried a pregnancy during that time. Abraham sent small remittances occasionally, though only to her. A year after they moved, Abraham visited, but he would speak only to Julia. He told her if she were ever in need she could always find a home with him in Santa Fe, but "on all occasions when Arthur's name was mentioned, she said, his father spoke most bitterly of him, called him a fool, a thief and a liar . . . and told the witness that she was a fool to stick to him as she did." By all accounts, Abraham was quite fond of young Julia. It was Arthur he couldn't abide.

Ilene

◇◇

I WAS INVITED TO a party at which a friend had arranged for ten-minute psychic sessions for all the guests. I don't typically go to parties that include psychics—this was, in fact, a first. And though I felt I had probably done enough divining by now, this session was free, so I went into my friend's office and sat down on the futon. This psychic's name was Ilene. She had tawny skin and vivid purple-red hair, hoop earrings and bluey-purple fingernails. She told me to keep my feet on the ground and keep my eyes open, lest I be taken over by a spirit. We had only ten minutes, so I got to the point: I asked about Julia.

"There was a boy," Ilene said. Julia was looking for a boy she had lost, a child. Was it the baby? I asked her. No, not the baby. "She found the baby right away." Ilene saw Julia pacing, wringing her hands, distraught. She needed to find her boy. He was the sweet one, the sensitive one. I began to list Julia's boys in order of birth: Paul, the invalid—"That was a birth injury," she told me—but no, it wasn't him. Arthur? Not him. And before I even said the name, I knew, in the same way I had known when my friend pulled the four of clubs. It was Julius. "That's him," Ilene said—she knew it had something to do with Julia's name, with jewels. Julia can't find him. He wasn't there when she died; she's waiting, still waiting, to see him. She doesn't know she's passed, she can't understand why there are people in her house. Ilene got teary-eyed. Julia had picked me, she said, because I could tell her story. "She wants people to know she wasn't crazy." Ilene saw Julia visiting me while I was in bed, stroking my hair. She saw lights, lots of lights. Our ten minutes were up.

I took Ilene's card and arranged a second visit in her office, a square room in Boulder's tallest office building, with a view from her window over the downtown rooftops. Ilene was also a life coach and a hypnotherapist—certificates lined the walls. She said she didn't remember much from our session; she tried to wipe these things from her mind when they were over. "She's south of here, yes?" she asked. "Not Mexico but something like that?" She saw a big house. The name Ida came to her. She told me I had an entourage of spirits around me—one of them was very tall. A child came forward; Ilene saw the letter J. "Definitely a very strong J," she said. The letter A came through as well. "Does that mean anything? It's very big. She's showing me a very large A, a capital A." A man, very protective. "She wasn't a very large woman, she was a little thing. He was a large man." I'd been thinking Abraham. Maybe Archbishop?

In Ilene's vision, Julia was wearing clothing typical of the period—a long skirt, cinched waist, high collar, sleeves. Ilene saw the "loss of a child," an infant, she thought, and its devastating effect on Julia—Julia wringing her hands again, pacing, putting her hands over her face. But Ilene's focus returned quickly to the other children, and in particular, to the little boy. "It's almost to the point where the other children that I see are kind of just standing to the side, she's put up this frozen barrier in her heart, she wants to love her children, but she's depleted of emotion. She's upset that she's not being mother to her other children." It must be the saddest thing, to forsake your children. The little boy kept coming up to her, but she had nothing to give him. Julius. She couldn't even help herself.

Ilene saw a house. She saw a long stairway, a grand staircase. She saw rooms with numbers on them, people sitting around the table—guests, not family. It looked like a boardinghouse. She saw a big porch. Kids running up and down the stairs. Servants. She saw burnt timbers at the top of the home, an outline of black charred beams. There was a fire. Julia was still there, pacing, anguished. "She can't let go. She can't leave." She saw Julia going up and down the staircase, pacing on the front porch, in her room, in a chair, a rocking chair where she swayed incessantly, closed off and shrunken down. Her spirit died in that chair, Ilene said, though her body may have expired elsewhere. "She let the heaviness consume her."

The heaviness was beginning to weigh on Ilene as well; she needed to let Julia go. I was thanking Ilene when my shoulder itched—a bug, or a stray hair. I batted it away. "That was Julia touching you," Ilene said.

twenty-eight

❧ MAP OF THE WORLD ❧

Bertha in her later years.

In 1919, Franz Kafka wrote a long—an excruciatingly long—letter to his father. It ran about sixty-five pages, each paragraph alternately scathing and self-pitying, seeping self- and paternal-loathing, oozing angst. He never mailed it; it was, rather, a letter to himself. A letter, he once said, was "a communication with ghosts"—with our memories and demons and fantasies. In Kafka's case, the ghost of his father loomed large. "You once asked me recently why I claim to be afraid of you," Franz began. "I did not know, as usual, what to answer, partly out of my fear of you and partly because the cause of this fear consists of too many details for me to put even halfway into words." Kafka blamed his father's "strength, noise and violent temper," for his own "weakly, timid, hesitant" constitution. He blamed his father for everything.

Kafka's father, Hermann, was, like Abraham, a German Jewish businessman, born a poor butcher's son in southern Bohemia about a decade after Abraham, and sent out to work at an early age. He eventually opened a haberdashery, which he transformed into a large and successful wholesale firm. He, too, married a woman named Julia, daughter of a wealthy merchant. Franz was their eldest son—their only surviving son—and all the expectations of the family weighed on his narrow shoulders. Like Abraham, Hermann was confident and overbearing; unlike Abraham, he was physically imposing, too. Hermann towered well over six feet; Franz, though nearly as tall, was stooped. Hermann was strong; Franz was consumptive. Hermann was a carnivore, a consumer; Franz declared himself a vegetarian. Hermann was a man of business; his son a tortured artist. "You had worked your way up so far on nothing but your own strength, consequently you had unlimited confidence in your own opinion," Kafka wrote. Franz, by contrast, had confidence in nothing. "From your armchair you ruled the world," Kafka wrote. "Your opinion was right, every other was mad, eccentric, *meshugge*, abnormal."

I read Kafka's letter while I was in Europe following the paths of Julia's family, and I couldn't help but think of Abraham, the blustery businessman, and his tortured sons, who were given all the opportunities—wealth, education—that their father didn't have as a child, and who found them oppressive. Kafka fled his father's shop. In the end, Arthur found an inelegant way to do the same. Betty Mae, my grandfather's cousin, had described Abraham as a "typical Germanic person. When he said, 'This is it,' there was no argument."

In Hebrew, Abraham means "father of a multitude"—Abraham, the biblical patriarch, brooked no rebellion. Men like Abraham Staab and Hermann Kafka stretched themselves far beyond their early prospects: they created themselves. But they found it difficult to understand such impulses in their own sons. Our children always grow to live in a foreign country, removed not necessarily by ships across the sea but by era and disposition. The German Jewish patriarchs, for all their wit and guile, could not intuit their children's needs. Hermann Kafka never actually beat his son. "But that shouting, the way you turned red in the face, the hasty undoing of your suspenders, laying them ready over the back of the chair, was almost worse for me," Franz wrote.

Kafka felt the burden of his father's expectations so strongly that, he wrote, his back became bent, which led first to weakness, and then to indigestion, "and with that the way was open to every form of hypochondria until finally, under the superhuman effort of wanting to marry . . . blood came from my lung." In their sons' quaking eyes, these German fathers swelled into giants—flickering, fearsome projections that obscured the whole sky. "Sometimes I imagine the map of the world spread out and you stretched diagonally across it," Kafka wrote to his father. Abraham stretched across his own proprietary expanse of desert, ruling his arid New World with a wagon and a ledger, controlling his offspring with prohibitions and manipulations, governing the mountains and mesas and arroyos and barren plains in which they dwelled.

But Julia cast a shadow as well. She had her own sad center of gravity—and perhaps it pulled her sons in. Julius had been the achiever, the stable one, it seemed. Then came the problems at school—he left Harvard Law after a year, I learned, and never graduated from Columbia, either—and then the stomachaches, the insomnia, and the suicide. There was Paul, nervous debility made flesh. And there was Arthur, whose resentment toward his family approached pathology.

Teddy seemed the only son to escape this curse. My father and his siblings remembered him well. Teddy was, Lizzie wrote, "the perfect dinner guest. He was handsome (although short), witty, could converse in three languages on almost any subject, and could make himself charming to both young and old." He was a man of impeccable taste.

After Julius died, Teddy quit his medical practice and traveled the world collecting art, which he kept in the high-ceilinged apartment on Madison Avenue where he spent winters—Renaissance altar cloths, Spanish Madonnas, Whistlers, Rembrandt etchings. He wore a houndstooth suit and a monocle (the *New Mexican* described him as "Chesterfieldian in demeanor and dress") and he had an impish sense of humor: everyone remembered with fondness his puns and malapropisms ("comparisons are odoriferous"; "don't cast asparagus"), his rhymes, his passion for tennis and swimming—he swam a mile every day well into old age. He was also gay ("Of course you know he was a homo," Betty Mae told me—and I did know that, though we of the younger generations used different language). He lived for many years with a man named Mr. McRae, a Presbyterian, teetotaling bookkeeper for the family business, who played gin rummy with Teddy at night.

But Teddy, like Julius and like Julia, suffered jags of severe depression. He was regularly coaxed by his family into lengthy stays at sanitariums, and he had a problem with prescription drugs—the problem being, specifically, that he had kept his medical license after he gave up his practice, and as late as the 1960s, when he was in his eighties and nineties, would write himself prescriptions for painkillers and sleeping pills. He also liked to partake of a pink alcohol-and-narcotic concoction called Excelsior Liniment, a patent medicine used "for the cure of rheumatism, neuralgia, pneumonia, pain in the back and side, lameness, headache, toothache, poison, sprains, bruises, burns, wounds, frost-bites, fractures, dislocations, ulcers, enlarged joints . . ." And also, perhaps, a broken heart.

Teddy lived in his last years at the Park Plaza, a high-rise apartment building in Albuquerque built in the fortresslike Brutalist style, which must have once appeared imposing and modern against Albuquerque's squat skyline—it is still the highest residential structure in the state— but now looks rather like a tenement. It was considered an upmarket address, and the well-heeled clientele probably frowned on the sight of Uncle Teddy prancing around the lobby, wizened, sloshed, and naked. The doorman would call Lizzie or my grandfather or Betty Mae, and they would bring him to the hospital to sober up. Then they would riffle through his cabinets to find the pills, liniments, and salves and throw them away. They eventually went to every pharmacy in town to warn about Teddy's drug problem, but he still managed to get what he wanted. Finally, they wrote to the state medical board, demanding that it revoke his license. "We were forever rushing him to the hospital after an overdose or fall," Lizzie wrote.

Julia's girls—Anna, Delia, and Bertha—seemed to fare better. They didn't labor under the same expectations as their brothers. Or perhaps they learned a lesson from caring for Julia in those sad last years: it was no way to live, shut in your room with the curtains closed, turned away from the world. They were all active and community- and charity-minded, engaged with the future. They were modern American women—and in that way, they moved beyond their mother.

Bertha became a tireless advocate for the state's youth, serving in the state Department of Public Welfare under three governors, taking in unwanted and orphaned children until they could be placed in foster homes, and campaigning for the first child labor law in New Mexico. She was almost Flora Spiegelberg-like in her energy and dedication. In photos from her later years, she looks cheerful; you see little of the uncertain, boy-crazy *Fräulein* from the travel journals. She grew ample, sure of herself—there seemed a generosity to her, and a warmth. The young woman from her journals must have seemed a ghost to her, as my own early self does to me—that frantic and eternally-put-upon twenty-four-year-old who first wrote about Julia's fate. Sometimes she visits me now, a specter spouting self-pity and wearing platform shoes, and I wonder how she and I could have occupied the same body.

Unlike her brothers, Bertha refused to allow the ghosts of her younger years to disturb her later ones. She would raise children and live a productive life under another man's roof—a kind man, by all accounts, and easier on his children than her father.

When Bertha married my great-grandfather Max, she deleted the word "obey" from her marriage vows. She had had enough obedience for one lifetime.

◇◇

The testimony in Arthur's trial wound down in early November 1914. There were, for a time, two holdouts on the jury, but in the end it upheld Julius's will. Arthur would get no more money. The family was stoic when the foreman read the verdict. "No tell-tale sign of the feeling with which the verdict was received appeared on the faces of any of those vitally concerned," wrote the *Journal*, "except that Arthur Staab smiled gamely, but the smile had the appearance of being forced."

Arthur and Julia went back to Oklahoma. He filed for a new trial, which was denied. It appears that he closed the laundry—he wasn't a popular laundryman, treating his workers perhaps worse than his father had treated him. One of his drivers beat him severely after an argument, and when his workers went on strike, the other laundries in town refused to help him with the extra work.

In 1920, Arthur and Julia traveled to Australia; I don't know why. They turned up in Los Angeles a decade later, and registered to vote there in 1932. They stayed, opening a curio shop in the Ambassador Hotel that sold Indian relics and art. There weren't any children.

My grandfather and Aunt Lizzie hadn't known of Arthur's existence until they made a trip to Los Angeles when they were teenagers, and Bertha "suddenly announced that we were going to call on our Uncle Arthur and Aunt Julia," Lizzie wrote. The shop can't have been very successful: Bertha sent money to support Arthur over the years, Lizzie said. But he was, anyway, his own man, living a life he chose, married to a woman he loved. "He was happy out there," Betty Mae told me.

twenty-nine

❧ AT FAMOUS LA POSADA ❧

The Staab House, shortly before the third story burned in 1924.

After Arthur's trial concluded, the news stories about the family trailed off. There were no Staabs left in Santa Fe. Julia, Abraham, and Julius were dead; the girls had married out of the name and moved to larger cities; Paul had relocated to Albuquerque to be closer to his sisters; Arthur was banished; Teddy, Julius's fortune in hand, was off collecting art. Santa Fe still held the state capital, but everything else was happening in Albuquerque. It was a different kind of town—sprawling across the Rio Grande, hemmed in by nothing, ruled by no one. If the Staab name appeared in the newspaper, it was now in the service of nostalgia, tidbits in the Twenty Years Ago column. This is how we fade from the world.

In 1920, Teddy and the sisters put the house on the market. "For Sale," the *New Mexican* announced, "The 'Staab' Mansion on Palace Avenue. Beautiful fifteen room residence with all modern conveniences, two story brick garage and living quarters, summer-house, fruit and shade trees, lawn, iron fence and paved street. Could not be duplicated for $50,000. Priced for Quick Sale, $20,000." It was still on the market in July 1921, now at a "bargain." It sold later that year to a man named L. E. Elliot, who planned to convert it into a high-class apartment building "or family hotel where wealthy tourists may find a delightful home to spend weeks or months in Santa Fe." People didn't know much about the place anymore: it was, the paper reported forty years after its construction, "built perhaps 20 years ago." Elliot turned it into a boardinghouse; he painted the red brick a cream color, the ironwork gray, and opened Julia's home to the world.

Early one morning three years later, a neighbor woke abruptly at two in the morning when his pony kicked down its stable doors in fright. Smoke was billowing from the back of the Staab home. The entire fire department—both engines—rushed to the scene. The flames had spread through the second story following the electrical wires—the conflagration had started in a switch box. It had crept up to the third story through the iron-sided air chambers between the original home and an annex added to the back, and burst out of the walls with sudden violence—exploding like a grenade, the newspaper reported. The city's former fire chief entered a room on the third floor and, seeing no signs of fire or smoke, opened a drawer in a built-in cupboard on the wall. A belch of flames shot out like an explosion, knocking him across the room and singeing his hair and eyebrows.

It was a "stubborn fire," the fire chief told the *New Mexican*. It took dozens of firefighters until morning to bring the flames under control. For more than six hours, firemen doused the structure with thousands of gallons of water, which poured through the ceilings to the lower floor. They pulled out what furniture they could and piled it on the lawns— walnut mirrors, bureaus, and chairs, much of it original furniture sold with the house. But other pieces remained inside. "A grand piano in the parlor to the left of the entrance was too heavy to move and it stood there," reported the paper the day after, "resounding to the patter of drops falling from the water-soaked ceiling." Oriental carpets, lace and damask curtains, tapestry-covered divans, Venetian glass chandeliers were all ruined. "Only the charred remains of much of this finery was visible to the eye today." The house suffered its own death.

And then, a resurrection of sorts. The third story was gone, entirely, and the newspaper speculated that Elliot wouldn't be able to keep the second story intact, either—that it would become a one-story dwelling like all the others in New Mexico, its grandeur flattened to desert scale. But Elliot managed to restore the second floor in the front part of the house. Julia's room, and Abraham's, next door, survive today.

Elliot held on to the home until the Great Depression; he lost it to foreclosure in 1934. It was owned by a bank for three years until a buyer was found: R. H. Nason, a collector of Southwestern art. Nason plastered stucco over Abraham's painstakingly assembled brick walls— that native mud that Abraham had worked so hard to surmount. He softened Abraham's straight lines, built a batch of small adobe casitas in the gardens, and turned the place into a motor lodge—La Posada, place of rest. The lodge hosted a summer arts school and retreat. There were two dance studios, along with facilities for tennis, swimming, badminton, and archery. Weddings—of strangers—were held there.

Santa Fe was becoming a place people visited for a holiday, to take in the old Spanish and Indian folkways and absorb the great American Southwest. There was no room on the premises for the memory of merchant princes or their wives.

The family who had once lived there also passed from the scene. Paul, the epileptic son, had died in early 1915, a few months after Arthur's verdict; Anna died in 1929 at age sixty-two; Bertha in 1933, at sixty-three, of a heart attack during an afternoon nap at her summer home in the mountains where I first read Lizzie's book and Bertha's diary.

Arthur lived another couple of decades—he died in Los Angeles in 1952, at the age of seventy-nine. There was no mention, in New Mexico, of his passing. And Delia lived into her eighties. She moved with her husband to Boston, where he worked for a firm that turned Southwestern wool into Eastern money. The family owned a large home in Brookline, and another at Manchester-by-the-Sea. Delia's gardens at the coast were splendid—she had inherited Julia's passion for gardening. When young relatives visited, Delia set them to work deadheading her rhododendrons—every day, no exceptions.

She was exacting, relatives told me, short and wide and fearsome. The bathroom towels in both of Delia's homes were changed twice a day, and each morning she lined up her maids in their black dresses and white aprons and gloves, put on white gloves of her own, and ran her fingers along the baseboards and molding, hunting for stray specks of dust.

After her husband died in 1935, Delia moved into the Braemore, a fashionable hotel in Boston, visiting New Mexico from time to time in the summer—there was always a big to-do when she arrived. But she was in her Boston hotel room when, on a December Saturday in 1951, she was found dead by her maid. It was not a gentle taking of leave. "The nude body of an elderly, wealthy widow, her throat slashed, was found Saturday in the bathroom of her suite in Hotel Braemore," reported a Montana newspaper. "Police said a knife was found near the body." She was facedown in a pool of blood.

I had assumed, after reading the newspaper account, that Delia had been murdered. In her book, Lizzie wrote that Delia had been stabbed in the "bosom." But then I had lunch with her grandson, a literary agent named Tom Wallace who lived in New York. We convened at the Century Association, a bookish private club full of leather furniture, floor-to-ceiling bookcases, and besuited New Yorkers—an uptown crowd. Tom was a member. He was in his late seventies, his hair only slightly grayed, his eyes dark-rimmed. He wore a gray flannel suit. His voice was lettered, each syllable distinctly pronounced, each vowel drawn out, a hint of New York in his A's and O's. As our drinks arrived, he told me, as others had, of Delia's flower gardens and the maids in their black-and-white uniforms, and of Delia checking for dust—for specks of disorder, Abraham's and Julia's Germanic rigidity still extant. Delia's table was always formal, he said: lobster thermidor, floating islands. She was philanthropic. Louis Brandeis, the Supreme Court justice, had been her lawyer. And Tom was certain, he said, as our sandwiches arrived and crowded our small oak table, that Delia hadn't been murdered.

Rather, she had committed suicide, just as her brother Julius had—and just as Tom's mother had, jumping from the window of her seventh-story apartment in Manhattan. He was convinced that Julia had also died this way. "That's what we were told," he said. The ice lurched in my Arnold Palmer.

The women in his family killed themselves. Men, too. It was a pattern, Tom believed—this was what Arthur had tried to explain in his lawsuit, about the "taint of insanity" that ran through the family. The psychologists call it "normalization": once a family has experienced suicide, it tends to happen again. It is a more insidious sort of inheritance. Julia taught Delia, who taught her own daughter—who was also named Julia.

So Tom and his family believed that his great-grandmother Julia had run up against her own despair and committed suicide—he didn't speculate on how she might have done it—and that Delia had, too. The parallels weren't specific: Julia had died in middle age, while Delia had outlived her husband and her once robust health. She had lost money to a Ponzi scheme after the war and was suffering the afflictions of old age. She was losing her sight, along with her money. So she decided to cut her neck.

Later, I searched old editions of Boston newspapers from 1951, and discovered, in the *Boston Traveler*, that "Mrs. Baer had been under treatment recently for a nervous condition." In the *Boston American*, I learned that the coroner had ruled Delia's wounds "self-inflicted and consistent with suicide."

It seemed a powerful exertion for an eighty-three-year-old woman—cutting one's own neck. But perhaps she, too, was haunted.

◇◇

Teddy—our pun-loving, art-collecting, nude-cavorting gay uncle—was the last to go. He died at the age of ninety-three in 1968, about three months after I was born—our final link to Julia. In Jewish tradition, it is said that we die twice. Once when we take our last breath, and again the last time somebody speaks our name.

For many years, the family continued to appear in legal notices in the papers, named in lawsuits against Abraham's heirs for his role in New Mexico real estate grabs fifty, a hundred years before. The cases went on for decades: eighty years after Julia's death, her dead children were still being named as defendants. Finally, the lawsuits went away, too. There was no one left to sue.

Soul piled on soul in that harsh country of serial conquest—Indian,

Spanish, Anglo, Jew. Story piled upon story: land taken and sold away, lives interrupted and truths lost, faith and superstition and sun and shadow.

◇◇

It was not until a few years after Teddy's death that there was any mention of a ghost at the Staab home. Not until 1975 did the newspapers discuss anything of a spectral nature in Julia's house, and when first they did, there was no indication whose ghost, exactly, it might be. "Up Palace Avenue a way at the famous La Posada, once the home of the prominent Staab family, the dining room . . . is said to be haunted," wrote the *New Mexican*.

> *Maids and waitresses never see the ghost there, but sometimes, when they are setting up or clearing after a private party or club meeting, they are puzzled or frightened. There is the sound of wind, as soft as a summer breeze, but as cold and cutting as the worst blizzard. It raises goose bumps and hair on the nape of the neck. Then it is gone. If decorative candles have been left burning they go out and a few wisps of smoke rise slowly, naturally and disappear. A long ago hostess or housekeeper anxious that everything be exactly right? No one seems to know. No sightings, no sounds save the soft whisper of wind and the intense momentary cold.*

The article also suggested that the ghost might be that of a devoted "Negro servant" named Ida. Ilene, the psychic I'd met at the party in Boulder, had mentioned an Ida. My gut welled up again, but I simply didn't know what to do with that sort of—what, evidence? Coincidence? Or simply the repetition of a common Victorian name—Julia, Emilie, Henriette, Ida?

It is probably no coincidence, though, that a ghost—whatever her name, whatever her station—appeared just as the mansion became truly timeworn. It was almost a hundred years old now—a relic, by American standards. In those hundred years, Victorian tastes had gone from fashionable to fusty and back to fashionable again. The year before La Posada's ghost first made the news, the hotel had undergone a renovation, returning the house to "the elegance of the 1880s," as the hotel's new advertisements put it. The old fireplaces, rosettes, and parquet floors were uncovered; the restaurant was renamed the "Staab Room" and equipped "for dining in the Victorian manner" (though true to its new era, it also featured "Santa Fe's most complete salad bar"). The past Julia once inhabited was now an appealing place to visit, her era far enough away to render it exotic. The mansion belonged now to historical, rather than personal, memory. Perhaps the renovation was a "trigger," as the ghost hunters put it, that brought Julia back into her lived world. But my gut—which had grown busy lately—told me that, more likely, the place was simply primed for a Victorian ghost story, its staff and visitors ready now to examine the house's past.

It was 1979 when a newspaper story first identified Julia as the hotel's spirit: "Julie Staab Still Watches Over Her Home," read the headline in the *Santa Fe Reporter*. It told of Alan Day, the employee who was mopping in one of the parlors when he saw Julia standing by the fireplace with her aura of sadness. The story hit all the dramatic highlights of Julia's life in the house: the "endless" social engagements in the home's yellow silk drawing room, the child who died as a baby, the rumors of Julia's insanity, the charge that she was locked in her room during the last years of her life. It quoted a woman named Consuelo Chavez Summers, who had been a little girl when Julia died, and who remembered Julia's last years as "quite mysterious." "Everyone knew that she had disappeared," Ms. Summers said.

It was now a full-blown ghost story with all the conventions—the shut-in, the disappearance, the woman undone. And it wasn't, I realized, entirely far from the truth of Julia's life. She *had* entertained in that home, and lost a child, and disappeared from the world. Only later did the story travel further from the truth, when it ventured into the larger world of ghost tours and television shows and the Internet. The variants of the story I heard as a teenager and the one I wrote about as a young woman began to appear. Julia became an alluring fair-skinned beauty, Abraham's "arm candy, " who hanged herself from the chandelier or was murdered by her husband. Her hair went white; she was chained to the radiator; she was a victim of spousal abuse. The tale took on the concerns and language of the day: battered wives, tyrannical husbands. As her fame as a ghost grew, the details shifted. Her ghost wore a red dress; a black one; all white. She was angry; she was hospitable; she was sad.

But no one knew for sure: Julia didn't communicate. She just appeared, a diaphanous specter, as insubstantial and fleeting as the record of her life, or a "draft of cool, stale air," as another article described her presence: vague, colorless, incomplete; a whisper from the past.

Jonathan

◇◇

I VISITED JONATHAN IN his adobe house in the shrubby foot-hills north of Santa Fe. He looked to be in his sixties. He had had a stroke the previous August, and he walked unsteadily with a four-footed cane. When he sat, he draped his paralyzed left hand carefully in his lap. His body had an uninhabited look about it, as if he had lost a great deal of weight.

Jonathan had worked as concierge at La Posada for three and a half years starting in 2002. He'd been anxious about taking the job, he said, because he had heard there was a ghost there, and he was "sensitive." He had known this since he was eight or nine, maybe ten, when he woke up in the middle of the night in his fam-ily's New York apartment, screaming, 'There's someone in the house! There's someone in the house!" His father told him to be quiet or suffer a spanking. His father gave vigorous spankings, so the next time it happened Jonathan didn't holler; he just watched as a dark figure walked around the corner into the bedroom he shared with his younger brother, past the large dresser, and then approached the beds. It was dark in the room, and the figure was even darker. It absorbed all the light, "yet I could describe to you exactly everything that this figure had on." A work shirt, a hat, a cane. When Jonathan told his mother the next morning, she showed him a photo of his dead grandfather. It was the same man.

That was Jonathan's first ghost. The second lived in his col-lege fraternity house, a land spirit that fretted over his safety. There was a dead woman who sat sewing at a bay window in his rental apartment in San Francisco, and an angry Indian warrior whom he could feel lurking outside the Palace of the Governors in Santa Fe when he walked by. When he was more able-bodied, Jonathan told me, he would always cross to the other side of the street to avoid it.

After he went to work at La Posada, Jonathan steered clear of the second floor, "for obvious reasons." But finally he was asked to take a group of guests to see Julia's room. Halfway up the stairs, the feeling hit him. "I'm breaking out into a sweat and I'm holding on to a banister as I'm going up," he said, "As I'm talking I'm getting dizzier and dizzier and feeling the sweat pouring down me." His knees buckled, and he had to put his hand behind himself to keep from sliding down the wall. When he went back downstairs, he felt instantly better. Twice more that week, he had to take guests up the stairs, and both times, he felt overcome.

Soon after, he received a visit. He was sitting at the concierge desk on the ground floor when he felt someone come up behind him as if walking fast—"their breeze." He saw someone standing just outside his peripheral vision, wearing black, but when he turned to look, there was nobody there. It happened again the next day; he felt as if someone were going to tap him on the shoulder. It happened again the day after. And then he knew. He walked to the foot of the old home's staircase. "Julia," he said, "I know that was you. I bothered you three times without telling you that I was coming up. I bothered you three times; you bothered me three times—touché."

Jonathan offered Julia a deal. "This is what I'm going to do. Every time I go up to show someone your room I will always let you know that I'm coming, because obviously it bothered you." He promised that as long as he was in her house, he would also wear a pin in his lapel, a mother-of-pearl rose, because Julia loved roses. They had an agreement.

Julia communicated with him in different ways over the years.

Sometimes when he got to the top of the stairs, a feeling of euphoria would come over him, "like a curtain coming down, and I knew it was her arms embracing me"—a "spirit hug," he called it. Sometimes when he spoke about Julia to coworkers or guests, lamps would flicker. Once, he gave a tour to a six-year-old boy who had brought Julia a rose—a wilted supermarket stem. They left it drooping limply in a Coke bottle in Julia's room. The next morning, the boy visited the room again, and the flower had revived. "I've never seen a more gorgeous crimson flower in my life," Jonathan said. "It looked like it had been freshly picked." Julia loved children, Jonathan explained. They had been her life.

Only once did Jonathan see Julia's face. It was Halloween, and ghost tours had been coming in all night. Jonathan sensed that Julia was feeling anxious. He walked up the stairs and sat down in the sitting area on the landing of the second floor. "I know you're upset," he told her, and he saw a reflection in the hallway's front windows. "There she was, standing in the corner wearing a Victorian cloak. Her hair was pure white. A feeling of euphoria came over him. "That was the first and only time I've seen her," he said.

But they spoke. She told him about Abraham. He had many lovers, she told him. "Of course she was not happy about his paramours—but she realized that there was nothing that she could do, and that's why her children became everything." I asked him how this particular conversation came about. "That's my secret," he told me. He smiled. People said she was crazy; Jonathan defended her. "She was a victim of her times," he said. "She had no choice."

Regardless, Julia was, in the end, too much for Jonathan. The hotel planned another renovation, and Jonathan started "feeling weird." He was getting angry all the time, especially at the decorators. And then one Monday the director of operations came in with a group of interior designers. "All I saw was red, I got so angry." He got up from his desk and charged at them, yelling about the changes they planned to make to the house.

This wasn't Jonathan; these weren't his words. He felt removed from himself, as if possessed. Later, he went to the staircase and spoke with Julia. "I've got to separate myself from you," he told her. "We're too close, this cannot happen again." The next day he got a call from another hotel offering him an interview for a concierge position. By Friday, he had tendered his resignation. He visited occasionally after that, and always wore his rose pin and said hello, but that was the end of it.

I asked Jonathan why he thought Julia stuck around. "Because," he said, "I think later in her life, that was the only thing she had left: the house." There was nothing else.

thirty

❧ THE RECORD OF WHAT WAS ❧

Room 100, the Julia Staab Suite.

There was nothing left for me, either. I had combed the deserts, the *Westfälische* riverbanks, the archives, and the Internet, prodding the dead, the dead prodding me. I had read all I could find, summoned every newspaper article and history book, interviewed every aging relative, consulted a boundless stream of psychics. Now there was only the house. I had walked through it so many times before, not realizing that every rosette and fixture and plaster molding had a dream behind it, a story that began in rural Germany and ended in the hands of strangers. It was time to visit Julia's room. To learn, once and at last, if the connection I had forged with Julia was real.

On a warm afternoon in late March, I swiped my key card and entered the "Julia Staab Suite"—room 100. The lock flashed green, the door creaking as it opened. I set down my bags and looked around. The room had changed since I'd seen it as a younger woman. It was bigger than I remembered—lighter, high-ceilinged, with a large four-poster bed made up cleanly in white linens. The walls were white, and also the curtains. Equinoctial sunshine streamed through the four arched windows. When I had visited before, the room had been stuffed with heavy furniture and dark drapery, the bed swathed in an intricate Victorian coverlet. There had been a rocker in the room—the one in which, perhaps, Julia had sought to comfort herself. That was gone. There had been a mirrored vanity, too, and a hairbrush that, the hotel staff told us, Julia had moved around from time to time. That had also disappeared.

The room now felt crisp and uncluttered, although there was an ornately carved settee and matching chairs that I was told had belonged to Julia, along with an elegant assortment of Victorian furniture that had not. Over the fireplace hung a portrait of a pale, dark-haired woman with rosebud lips. She wore a white gown, florets adorning her arms and décolletage. Many of the Internet postings I'd read suggested that the woman in the portrait was Julia, but this woman looked nothing like the photos that I'd seen, nor like any of the Staab girls, and I could see from the signature that that portrait had been painted in 1936, around the time the Nasons bought the house—another stranger in Julia's room.

A small writing desk sat in front of the windows that once had overlooked the home's front entrance and from where, I imagine, Julia could gaze out at the passing street life, the hustle-bustle of human hope. The view today is different. I could still see the mountains in the thickening dusk, but I couldn't see the gardens or the street below—just the flat roof of the adobe addition that housed the restaurant and reception area.

I poked around the room a bit, peering into the closets and under the bed. The bathroom harbored an air of slight decrepitude that even a luxury resort couldn't buff away: crumbling black-and-white tile beside the bathtub—probably not the same tub, but perhaps one located on the same spot where Julia had died. I felt the general scalp-prickle one experiences when thinking a ghost might be present; my heart rate ticked up a notch. I perched on the end of the four-poster bed and watched the mountains darken.

I had dinner on Water Street, around the corner from the spot where Abraham's empire had once been headquartered. I ordered a huge mound of soggy blue-corn enchiladas that Julia, with her continental palate, would probably have abhorred, then wandered back through the still-warm March night to the hotel bar that was once the family parlor. There was a jazz band entertaining a number of sleek people who looked like they were from Los Angeles, with sport jackets and gelled hair that crested above the scalp in stiff, calculated peaks, like meringues. I sat on a velvet love seat and ate lemon curd, observing those invaders in Julia's home, and then, when I had run out of excuses, I headed up to sleep in the room where my great-great-grandmother had died.

<><>

My night in Julia's room would be that simple: a night in Julia's room. I would brush my teeth, take off my clothes, and climb under the sheets. I would bring no equipment—EVP recordings and EMF monitors held no persuasive power for me. My two years of poking around in the world of ghosts had convinced me of one thing: no amount of evidence can persuade anyone not already inclined toward the supernatural that ghosts walk the earth, that a dead woman named Julia still visits the living.

That isn't to say I wasn't nervous—I was. I both wanted to feel something and feared that I would; wanted a visitation and knew it wouldn't satisfy me. I read for a while, then turned off the light. Sleep came, and then it went, interrupted by the noise from the bar below and the click of the heating system and the rhythms of my own unease in that place between sleep and waking where reason is no defense. I was in the center of that desert city, as Julia had been. Sounds carried far in the sharp air. I imagined her sitting in the room while Delia's wedding went on below, too consumed by the demons of her past to live in the present. Voices rose and fell, the jazz drums pattered, a saxophone warbled. I drifted off and woke. I got up and fetched a lighter blanket. I drifted off again—I know this because I dreamed that Julia was pulling the covers off me. I woke in a sweat, covers still on, and got up to turn on the room's fan.

In the daytime, I am an optimist. I can find a comforting explanation for every disturbing detail—for setbacks and weakness and failure and loss. I am, while the sun is up, a generally resilient and practical person. But at night, when sleep creeps in or fails to, I become an altogether less rational creature. I fret; I agonize; I assume the worst; I suffer terminal conditions. It is in those hours that I inhabit a world of darker designs. It is in those dark hours that Julia's despair makes perfect sense. In the depth of the night, I believe in unhappy endings. I believe in my ghosts.

I drank some water. I went to the bathroom once or three times. I looked out the window at the roof of the adobe structure below, its skylights and ductwork, open-ended shapes in the dark, subject to interpretation. I heard clatter and chatter, drums pulsing, heaters clunking, footsteps on the stairs, good-byes being said, locks sliding, doors opening, doors closing, and oddly, a dog barking in the front hallway, bar stools protesting, glasses clinking, floors creaking, walls settling, and then mercifully, silence. I slept.

Then, just before dawn, something happened—though it sounds less convincing to me each time I tell it. I had fallen asleep on my side. I awoke just before light, not with a start this time, but slowly. I opened my eyes, and on the wall in front of me, just above the doorway to the bathroom, were green lights. They were moving. I was barely awake—or perhaps still asleep, in a hypnopompic state—and I first thought that they were indicators on a burglar alarm or smoke detector. But they seemed to dance with no particular rhythm, or rather they danced with a rhythm all their own. They moved ever faster and more intently, and then, just as I was about to sit up and investigate, they turned red and orange, zipped off to the right, and disappeared.

Like that. There was nothing else. I didn't have time to run or to fall down the stairs, much less dismiss them.

The vision happened quickly and ended quickly, and I wasn't at all sure that I had seen something, except that I had. I wasn't wearing my glasses. I had spent the day thinking about ghosts—I had orbs on the brain. I didn't know if I was awake or asleep, and I couldn't be entirely certain that I hadn't in fact just seen flashers from a police cruiser on the street below, though the room's blackout curtains let in only a narrow crack of light from the street. Only when it grew light enough for me to see out did I recall that the view of the street was obscured completely by the adobe roof below; if the lights had come from a police car, they would have had to double reflect in some inexplicable way. Nor was there a burglar alarm or a smoke detector on the wall where I'd seen the lights. In those hazy early hours, it seemed completely reasonable that I had in fact seen orbs—a gracious sign from Julia, as if she had wanted to say hello: I'm here.

Make of it what you want.

Even as the moment happens, it's over. All we have left are blurry recollections, clouded by self and predisposition and the limits of the human brain. I didn't sleep any more that early dawn. I lay there and stared at the wall and thought about Julia. I didn't feel scared. I felt lucky. I felt touched—in the demonstrative sense, and perhaps in the crazy sense as well. My father would suggest later that perhaps I had seen the lights from a UFO. This is how my family contends with the inexplicable—we joke. Other families—Jonathan's, for instance—might take those signs more literally. And thus our beliefs are transmitted. Memory becomes news, the stories we share, sifted by the sharers; news becomes the record of what was. The story and the truth are not always the same. Sometimes they are.

I checked out later that morning, and the woman at the reception desk asked, of course, if I'd seen any ghosts. The higher the sun climbed in the sky, the sillier I felt. But I told her about the lights. And she told me that a woman in room 310, just above the spa in Julia's old gardens, had awakened in the night to lights in her room, too. They moved around on one wall as well, though in her case it was three in the morning, not five. They came and went quickly, like mine, and that woman, too, hadn't slept the rest of the night. She had already checked out. There would be no comparing of notes, just perception and suggestion and memory and myth, mingling in an indistinct swirl, like dust motes—like orbs.

thirty-one

❧ DUST ❧

The Staab plot at Fairview Cemetery.

When I was in Prague, I visited the old Jewish cemetery—a teeming jumble of waferlike stones pressed together in a walled-in courtyard. It was an astonishing sight: twelve layers of graves, fallen into each other as the earth heaved and sank, each generation more deeply buried by the settling of earth and history. There were so many souls there, their stones worn to anonymity, their names forgotten.

Yet we remember Julia. She was no monarch or saint or martyr. She left behind no art or letters, only children and a legend—and perhaps a spirit—that keep her from slipping underground for good. Her story persists, changing with each generation, each summoning, each novel, each website, each whispered exchange in the hallway. It changes with each teller.

Now it's mine to tell.

Julia Staab sailed from Germany to a life that didn't suit her. Her health, physical and mental, was fragile. Life's gusts tossed her around. She missed her home and the life she would have had in Germany among her sisters—she had no way of knowing how that land would one day turn on them. She found motherhood to be taxing, each child taking a part of her until there was little of her left. She lost her last baby—the child of the darkness. But the darkness was not in the baby, but in herself.

Julia had few sources of relief available to her: no community of sisters, no cognitive-behavioral therapists, no serotonin reuptake inhibitors. The husband who chose her may not have understood her, though I believe he tried. Perhaps he was a despot—he was certainly no angel—but if so, he was a despot of his era; no better or worse. He wanted to help Julia, but he couldn't. And so she faded slowly into herself, and away from her family and the world. She failed—mentally, physically, both—and she died, in her chamber, in her bathtub perhaps?

Suicide or no, the cause was the same: she could no longer abide living. Her sadness followed her from Germany to New Mexico and back, impervious to geography and companionship. Whether she was a victim of her times, or her husband, or her circumstances, or her religion, she was also, in the end, a victim of her own constitution. If her soul was divided in death, as Sarina had explained during our reading, I realize now it was because she was never whole in life.

She wanted to die, and yet her ghost story has kept her alive. She was invisible in all the years that she lived as flesh and blood—Mrs. A. Staab, an accessory first, and later, an invalid and recluse. She was rich by birth and marriage, but in her world, in her time, she was still barely there. Yet here she floats, buoyed by the tale of her adversity, while stronger souls have sunk into history. The philosopher Hannah Arendt once said that posthumous fame is the saddest sort. That is the bitter reward to which Julia has been consigned.

This is what I've come to understand about ghost stories: it's not so much the ghost that keeps the dead alive to us as it is the story. Ghost stories make visible the forgotten, the repressed, and the discarded. A hundred and twenty years after Julia was "called to her long rest," people still talk about her. They lead tours to her home and write books about her—this fragile woman on the rough frontier, far from her family of origin and trapped in a world not of her making. Abraham dragged her across the ocean and the desert. But in death, she is the one who exerts a pull. She has dragged me into the past and made me custodian of her story, tracing a ghost outline of her years on earth.

And I believe this particular story has something to teach us. Ghosts connect us both to memory and to the world we cannot fully know. It is an unseen world in which the questions are leading and the answers vague and often contradictory, taking us beyond the methods of research that I hold dear. But in attempting to visit that world, in asking the psychics the questions the archives couldn't answer about motives and emotions and secrets, I became aware that those intuitive and emotional truths lie at the heart of most of the stories we tell ourselves. It is the truths between the facts that tell us who we are.

I once thought of Julia's ghost as a joke and an anecdote. Now I consider it a gift. It has lured me into a world I would have never known. So of course I believe in ghosts. I believe in the power of the past. I believe that we can be haunted.

Julia has touched me, carrying a message about how we live and what we treasure and what we leave behind. It was not, in the end, Julia's ghost that taught me these things, but her life. Observing how the task of living wore on her, I see how the past can engulf us. We can absorb and become our losses, or we can accept them and try our hardest to face forward and go on living.

The evening after my first child was born, my parents and husband and I ate dinner on an outdoor balcony at the hospital. As we sat, the baby beside us in her bassinet, an owl flew onto the rail of the balcony. They say that owls are messengers from the other world. This one watched us in the gathering dusk, and we watched the owl and baby both, the air feathered with hope and memory, and we all felt the weight of the past—loved ones long gone, winging in to see the future arriving.

We came home, and one sunny afternoon a week or so after she was born—a perfect May afternoon when the sun streamed through the open windows and the air held our skin in equipoise, the penstemon fragrant and groping toward life—I felt the softness of my daughter's cheeks and lips and realized that there, swaddled and held close, was life beyond my own. She carried me forward; I linked her to the past. We named her Delia—for my husband's grandmother and also for Julia's daughter. There are Julias and Teddys and now Delias in our family, names that stretch across the generations, because the past can illuminate the future, and perhaps the future can also mend the past.

I am at an age now at which death lurks more obviously and takes more readily. The worst things we fear, the things that haunt us at night, are certain to happen—those we love will die, the body will decline, and then we too will die. Life flees like a shadow; it slips by like a field mouse. However we live—weak or strong, rich or poor—we leave dust; words and objects; stories and documents; brick mansions and yellowed photographs and letters of light on a screen. Traces of genetic code found in ever more distant generations: a twisted double helix. But also, the earth under our feet, the mountains that hang above us, clouds, rocks, eagles, vultures, scrub oak, piñon, apricots, burros, alleys, streets, fences, lightning, snow, sleep. We leave them behind. Julia was resilient in death. I prefer resilience in life.

◇◇

After I checked out of La Posada, I stopped at the cemetery where Julia and Abraham are buried. It must have once been a pastoral spot on the edge of the city, the resting place of Santa Fe's Anglo elite, but it had fallen on hard times. The once lush Kentucky bluegrass lawn was gone—the grass had died away when the cemetery's board could no longer pay its water bill. Prairie dogs had found favorable territory here, digging networks of holes and tunnels that had begun to undermine many of the gravestones. When the graveyard was a going concern, I read, the indigent had been buried in winding sheets or cardboard coffins, which had later disintegrated. Burrowing animals had begun to bring up bones from below.

I found no bones the day I visited Julia's grave; only earth so barren it could barely support weeds, the ground returned to desert, red-beige and prickly, the once grand American elms and lindens and horse chestnuts also bare—dead, I feared, for lack of water, though perhaps spring came later to this place of the dead than elsewhere. Outside the chain-link fence was a busy six-lane road crammed with rush-hour traffic, and an electrical substation, wires and superstructure, and a state building of the institutional sort—thick walls and few windows. There was also, incongruously, a preschool play-yard providing the only splash of color in the entire tableau, aside from a few fake flowers laid on a few newer graves.

I drove the perimeter of the graveyard looking for the Jewish section, which I assumed would be to the side or in the back, as it is in most cemeteries. I passed the ample mausoleum of Abraham's crony Thomas Catron (b. 1840, d. 1921)—a volume of marble, Doric columns, intricate stonework, and sheer cubic mass taking up lots of real estate, as Catron had in life. I circled around the edge looking for a telltale Star of David that might point me in the right direction, then cut through the center to look on the opposite side.

I was so focused on stars and Jewish last names that I almost passed the enormous, obelisk-like monument, which looked not much different from the one in the Santa Fe Plaza that celebrated the killing of "savage Indians." It sat dead center, a Gothic "S" adorning the top and the word STAAB carved in simpler letters below. Of course Abraham's plot wasn't in the Jewish section. There wasn't a Jewish section, because there weren't enough Jews in turn-of-the-twentieth-century Santa Fe to have a section—and besides, Abraham was always in the middle of things; and Julia, too, by default.

So their plot was at the cemetery's heart, their monument the tallest I could see—fifteen, twenty feet high—topped with what looked like an urn draped in cloth, and carved of pale gray granite. At the obelisk's foot was a square of dirt surrounded by a low granite wall. It must once have contained flowers or the grass that had blanketed the rest of the cemetery, but now it sheltered dirt and weeds. Surrounding the obelisk were five smaller gravestones and a wrought-iron fence, the only one in the cemetery—enclosures had been outlawed in 1903 to place an emphasis on "uncluttered dignity without ostentation." But Abraham made his own rules; the iron barrier set the family apart, as Abraham always had—by station, not religion. And there the family was, still gathered around Abraham, ordered and contained.

I yanked the wrought-iron gate and walked into the plot for a closer look. Carved in the marble curb at the monument's foot were the words FATHER on the left and MOTHER on the right. This was how they wanted to be remembered in the end. On the tower itself were Abraham's and Julia's dates: set in stone, impervious to rumor or myth. ABRAHAM STAAB, read the a simple, blocky font, FEB, 1839–JAN, 1913. And below, in letters more delicate, dates closer together, 1844–1896: JULIE STAAB.

They spelled it the old German way, because even after thirty years in America, she was still German.

The older girls had been buried in other cities with their husbands; the sons were all here. To the left of the monument lay Paul's grave, a granite rectangle that flared slightly at the top, adorned with the same Gothic "S" as the obelisk. A large tumbleweed had come to rest against the stone. Julius's grave flanked the spire to the right, the same shape as Paul's, matching granite. To Julius's left lay Teddy, under a modest polished marker, flat to the ground and covered in coarse red dirt and fire ants. Arthur's grave occupied a parallel spot on the left, the same black marble and pattern as Teddy's. Hadn't he been banished? Why wasn't he buried with his wife, his own Julia, for whom he had given up so much? More questions, the answers available only to the dead.

To Teddy's right, tucked away in the back corner, was the lost baby, Henriette. She had been given a small, sweet marker, white marble, a delicate squared-off arch. Its shape reminded me of Julia's bedroom windows. It was of a soft stone, the day of her death too eroded to decipher.

I took a few photos, and when there was nothing left to study, I pulled the tumbleweed from Paul's grave and threw it out into the dust beyond the fence. Henriette's stone was almost entirely obscured by a lilac shrub—a pretty little bush, the only living thing. The world moves on. They were dust, these ancestors of mine, ghosts: an aggregation of stories and dates, of fuzzy recollections and rhetorical questions, of faded photos and crumbling documents. I brushed the ants and earth from Teddy's stone so the sun could once again shine on his name—so it could be read.

Judith

◇◇

Her name was Judith. She talked to spirits using an L-shaped divining rod—she was a dowser. She had come recommended by a friend who had felt and seen ghosts in his home in Boulder's northern foothills, a heaving landscape of striated sandstone and ponderosa pine. He'd wakened one morning feeling that someone was sitting on his chest. He'd contacted Judith, who came to his home. His house and the ground under it were crowded, she'd said, with the ghosts of four Ute Indians and two miners, seventeen demonic spirits, three negative vortexes, six celestial holes and 102 ancient curses. She expelled the ghosts, closed the holes and vortexes, and erased the curses. She had the power to do this.

A ghost hunter told me once that there is no surefire way to exorcise a ghost. "If a ghost doesn't want to leave, it's not leaving," he said. He believed that spirits sometimes have a message to get across, and if you acknowledge them and listen, they will fall quiet. Other times, you simply have to let ghosts know that a place is not theirs anymore. In the Jewish kabbalah tradition, ghosts linger when the soul is tortured by unfinished business. By finishing that business, you lay them to rest.

"I think I met your great-great-grandmother about fifteen or sixteen years ago," Judith told me on the telephone when I called her. This had happened before Judith had become a dowser, and although she was sensitive to spirits back then, she had no way of communicating with them. She had gone on a trip to Santa Fe and had slept in Julia's room and felt uncomfortable—as if she was staying without permission. When she checked out, the clerk asked if she'd met Julia, and she realized that she had indeed felt a ghost.

I visited Judith at her house on a golf course subdivision east of my place on a frigid December morning. She made me tea, and we sat in her living room—carpeted, with full-length windows, the walls hung with New Mexican art and knickknacks: santos and rusted-sage landscapes, blue-green oil paintings of adobe churches. Judith had short auburn hair, a smattering of freckles, a few wrinkles. She wore a fleece vest and red down slippers. Dowsing, she told me, was an ancient art—Moses used dowsers in the desert to help him find water; the pioneers took dowsers with them on their wagon trains. The early divining rods were simple forked pieces of wood, shaped like a "Y," which would dip toward a line of water or energy.

Judith used a copper L-rod, which spun forward or backward in response to the questions she posed to her spirit guide: backward meant yes, forward, no. She had been using it before I came in. Julia, she told me, had been messing with her energy for three days, and Judith was feeling very dizzy and jittery. "She's in here with us right now," Judith said. "She's here because you're here—she wasn't here before you called me. I think there's something she wants to talk to you about."

Judith sat in an armchair with her eyes closed and the rod in her lap. When she asked a yes or no question, she raised the L-shaped rod in front of her. "Julia, please tell us, did you long for Germany?" It spun backward: yes. She asked if Julia always longed for Germany. Backward again. Did she at some times love Santa Fe? Yes. Did she love Abraham? Yes. Did she feel that Abraham loved her? Did she miss him when he went away? Yes, she did. The archbishop—were they friends? Yes. Lovers? "Well, I don't think they were lovers," Judith said after a complicated series of spins on the stick, "but I think they did have a love for one another. Of course I have to rely on what she's saying"—

spirit conversations being subjective, as I knew by now. "Maybe she's protecting him. Maybe she doesn't want to tell me."

Judith asked about the pregnancies, the miscarriages. She asked about the insanity, and Julia again insisted that despite all the rumors, she had not lost her mind. But she was sad. There was an accident. "I'm getting a thought that maybe she got pushed into harm's way. Did you get pushed into harm's way?" Yes. A fall. "I don't know how much of that is true," Judith added. After the accident, Julia was an invalid—physically, mentally. She was restrained in her room. She had mental breakdowns, but Abraham never placed her in an asylum. He wouldn't do that to her.

Judith stood up in her red down slippers and vest, held the rod higher, and asked Julia about her death. "Did you feel you were in a living hell?" Yes. "Did you no longer want to live like you were living?" Yes. These were leading questions, but they led us to answers we wanted. Judith also believed Julia had drowned in the tub, after taking something—poison, laudanum, too much of something. And then Julia stayed on—she chose to stay, Julia told Judith. Not because of Abraham. Not because of her children still there. Because of the house. It was her security, and she was afraid to move away from it. "Did you feel you did not deserve to go to heaven?" Judith asked. Yes. "Do you still feel that way? That you're not acceptable to God?" She didn't. "She's changed her point of view now," Judith told me.

"Is there something else you want us to know at this time?" Judith asked her. Julia said that she enjoyed being a mother. She loved her children. "And she loves you," Judith told me. Julia was glad I was telling her story.

And now she was ready to leave. "She wants peace," she said. "She wants to go where Abraham is, and her children. She's asking, Will you please take me home? She said, I'm begging you."

And then Judith started crying. She was sorry that she hadn't helped Julia when she'd first met her in the hotel so many years ago, and she asked if it was OK with me if she set Julia free from the limbo in which she was suffering. It was fine with me, more than fine, and there, as I sat in Judith's living room, talking to an ancestor through an L-shaped copper rod—an ancestor said to be confined in a strange halfway compartment of the afterworld where the dead dwell and the past still lives—my specific questions about Julia's life and death ceased to matter. I just wanted her to find a place of rest.

As the morning sun struggled to warm the frosted grass on the golf course, I closed my eyes and held Julia tightly. It was time to let her go. The dead are past our help, though it's hard, sometimes, to live without them. They take a piece of us with them when they leave, and we must learn to live reduced. We must live and let the dead be dead. "Look at me," Eurydice begged Orpheus, and he did. We look hard at those who have left us, and then we let them go—that is what it is to love them.

"There's an archangel I ask to take people home," Judith said. His name was Metatron—a Jewish archangel known as a "lesser Yahweh." He had three pairs of wings and he was enormous and powerful; he taught us, Judith explained, to let go gracefully of the things that tether us. Judith closed her eyes and asked the archangel Metatron to remove all pain and suffering from Julia, to surround her with love and joy.

This is what Julia has taught me: to embrace this world, this beautiful world, with every molecule I contain—muscle, tendon, blood, and brain—like my grandfather, like Wolfgang, like Flora Spiegelberg. And like Bertha, too, in the end. To seize it. And then to let it go.

Judith smiled, eyes still closed, tears streaking her cheeks. "I am feeling that it's like a celebration party, they're all waiting for her, they're all going, Yaaay! All your family is there for her, all of them who have passed on." Now Judith addressed her questions to Metatron. "Is Julia still earthbound?" she asked. The rod swung forward: no. "Did she rise up? Did she go home?" Yes, she did. She did. "I knew she went home," Judith told me. "I could feel the joy of her family. They were all waiting for her."

I want so much to believe it.

❧ ACKNOWLEDGMENTS ❧

Though I wrote *American Ghost* without any help from the spirits, as far as I know, this book was a communal effort—a testament to how our family stories keep the past from slipping entirely from our grasp. So I owe first and foremost a huge debt to those in my family who left paths for me to follow. Julia Schuster Staab and Bertha Staab Nordhaus were present in every page of this book. I hope they would approve.

I am lucky to come from a family of storytellers: especially my great-aunt Elizabeth Nordhaus Minces, whose collection of family memories propelled this book into being; my grandmother, Virginia Riggs Nordhaus, a talented writer who, in another generation, would have had more opportunities to express that talent; my grandfather, Robert J. Nordhaus, who never saw an ear he didn't want to bend; my father, Robert R. Nordhaus, who forgets nothing, ever; and my uncle and aunt Dick and Mary Nordhaus, who know how to cut straight to the drama.

I owe a special debt of gratitude to my aunt Betsy Messeca, who has been tireless in tracking our family's path through Europe to America. Without her laboriously archived photographs, maps, emails, and family trees, I wouldn't have known where to start. My father's cousin Nancy Paxton provided a number of family documents and photographs that also helped point me in the right direction. Nancy's late sister Judy Paynter kept diligent track of our family's legacy; her daughter Rhonda Paynter is caretaker of many important things, and I owe her huge for finding that diary.

Betty Mae Hartman, Don Wallace, Tom Wallace, Lee Meyer, and Wolfgang Mueller were all generous in sharing their time and stories, offering the view from other branches of the Staab and Schuster family trees. Wolfgang did not live long enough to read the manuscript; with his great gusto for being, I somehow thought he would live forever. Thanks also to Sonya Mueller for her help in finding photos, and to Felix Warburg for photos and tales of the Spiegelberg family.

I was fortunate to have found such generous and well-informed German historians to help me trace the Schuster family's path in the Old World. I can't offer enough gratitude to Margit Naarmann, who has spent a career chronicling the lives and difficult times of the Jewish families in her corner of Westphalia. I am also grateful for Manfred Willeke's extensive research on the history of the towns of Lügde and Bad Pyrmont, and his gallantry in squiring us through those worlds.

My German-language skills were woefully inadequate for such a German-intensive project. Jim Robinson, a masterly translator, helped me make up for those deficits. Many thanks also to Lynne Sullivan, Buzzy Jackson, and Megan Smolenyak Smolenyak for their help in guiding me through the mazelike world of family genealogy and DNA research. Willie Sutherland helped me sift through newspaper stories and provided bibliographical backup; young eyes are very helpful when engaging with old microfilm. Joanna Hershon, who wrote *The German Bride*, the evocative novel based on Julia Staab's life, offered unconditional support and information.

I am also grateful to John Lorenzen, Jonathan Mason, Judith Mangus, Juli Somers, Steve Hart, Sarina Baptista, Ilene Blum, Ed Conklin, Judy Cooper, Misha Johnson, Karl Pfeiffer, and Connor Randall for sharing their visions of the world beyond. My mother-in-law, Toni Barkett, is not only clairsentient, she is also courageous—game for both visiting psychic colonies and tending toddler grandchildren.

My travel companions on various ghost-hunting adventures, Monica Nordhaus and Emilia Noullet, did not get the credit in the book that they deserved. But wasn't it fun? Big thanks to Kristin Lepisto and the staff at La Posada for hosting me and letting me snoop around the place. Thanks also to the Stanley Hotel for its hospitality.

My monthly writers group provided such helpful feedback: Morgan Bazilian, Deborah Fryer, Buzzy Jackson, Carol Kauder, Radha Marcum, Michelle Theall, Rachel Walker, and Rachel Weaver. My *other* gang of writers—Hillary Rosner, Melanie Warner, and Florence Williams—pitched in with advice, support, and indispensable lunchtime getaways.

To those who muddled through various incarnations and pieces of this manuscript—Coralie Hunter, Buzzy Jackson, Bob Nordhaus, Mary Nordhaus, Ted Nordhaus, Rhonda Paynter, Myra Rich, Hilary Reyl, Hillary Rosner, and Melanie Warner—thank you, thank you. You made this a better book.

Carol Byerly—historian, friend, fellow stickler—read an especially early and troubled draft. I can't thank her enough for her merciless feedback. Rachel Walker, too, gave the manuscript the thorough and acute read it desperately needed. Meg Knox helped me channel my inner memoirist and aided in transforming the manuscript into something far more crisp and elegant.

Richard Pine, my agent, seemed to know exactly when to be encouraging and when to take me to task. Many thanks to him for his warmth and enthusiasm, and also to Eliza Rothstein for her input and help. Michael Signorelli took a chance on my first book and advocated for the second, then left me in great hands with Maya Ziv, who brought a fiction editor's eye to the book's pacing and plotting and kept me working at it until we were both happy.

And then there is my mother, Jean Nordhaus: traveling companion, translator, reader (of everything except maps), editor, counselor, and all-around word guru. How could I possibly write this book—any book—without you?

I always knew I was lucky to be married to Brent Barkett—companion, friend, child-whisperer, griller of rare meat, laundry folder, cycling domestique, voice of reason, and force of calm. But after spending three years thinking about life with nineteenth-century husbands, I find myself even more awed by and appreciative of all he is and does. Our children, Delia and Milo, bring me the future, keep me in the present, remind me of our connection to the past, and always make it fun. A world that contains these people is an infinitely better place.

❈ A NOTE ON SOURCES ❈

My research for *American Ghost* was vast and varied, ranging from old territorial newspapers to nineteenth-century travel journals, from historical monographs to biographies to Internet genealogy forums. But it began with one dusty, slim volume that I found in a leaded-glass bookshelf at my great-grandfather's mountain home in the eastern Sangre de Cristo Mountains. My great-aunt Elizabeth Nordhaus Minces wrote *The Family: Early Days in New Mexico* the year of her death in 1980. Without both the information and the inspiration contained within it, *American Ghost* wouldn't exist.

Nor would my family's story have been nearly as engaging for me without the great luck of finding the 1891–93 travel diary of my great-grandmother Bertha Staab Nordhaus, a scuffed, leather-bound notebook that surfaced in a neglected moving box just when I needed it, and opened up a world. My grandmother Virginia Nordhaus's reminiscences of her own years as a young bride in New Mexico, *Unsent Letters*, also helped me understand what it might have been like for "cultured" women from the East in the days when a move to New Mexico was akin to a journey to another planet. These family memories and documents are remarkable gifts; they tell us where we came from. I am fortunate that the women in my family felt compelled to write about their days and their memories, and thus saved them for the rest of us.

Supplementing those penned reminiscences were oral histories. There is no one left in my family who remembers Julia Staab, but I was able to interview a number of relatives, distant and close, who knew

Julia's children. My interviews with Betty Mae Hartman, Wolfgang Mueller, and Don and Tom Wallace provided essential insight into the Staab and Schuster family cultures. Other relatives—my father, Bob Nordhaus, uncles and aunts Dick and Mary Nordhaus and Betsy Messeca, along with my father's cousin Nancy Paxton—shared memories and gossip in more informal settings.

I found additional information on the Staab family in historical and online archives. The New Mexico Jewish Historical Society has compiled extensive files on New Mexico's early Jewish families, including the Staabs and the Spiegelbergs. Those files are kept within the New Mexico State Archives. Another batch of Staab folders resides in the Museum of New Mexico History Archive, where I also perused the papers of Julia Staab's physician, W. S. Harroun. The Bloom Southwest Jewish Archives at the University of Arizona contain the papers of Floyd Fierman, who conducted extensive research on Southwestern Jewish families, including the Staabs, in the middle of the last century.

I was lucky to embark on this research project in an era when so much that once languished in dusty and far-flung archives is now searchable and readable online. The depth and breadth of the newspaper archives now available on the Internet are truly a gift to any historical or genealogical researcher. I conducted my searches through GenealogyBank.com, NewspaperArchive.org, and Newspapers.com. Through those services, I found references to the Staabs in nearly forty newspapers across the United States. The *Santa Fe New Mexican*, the *Santa Fe Weekly Gazette*, and the *Albuquerque Journal* were particularly valuable resources in my search.

I also found a wealth of first-person accounts about the early days in territorial New Mexico. Josiah Gregg's *Commerce of the Prairies*, which recounts his journeys across the Santa Fe Trail in the 1830s, is a Western classic, masterfully written and spectacularly observant. Susan Shelby Magoffin's diaries of her 1846 journey to Santa Fe, *Down the Santa Fe Trail and into Mexico*, is the earliest known first-person account of that trail written by a woman, and it provides intimate detail of a female traveler's adventures and tribulations. Sister Blandina Segale's *At the End of the Santa Fe Trail*, a not-to-be-missed rendering of a young nun's days in territorial New Mexico, is full of life and sass and magnificent detail. The archaeologist Adolph Bandelier's four-volume *Southwestern Journals* also provide insight into the conditions and social scene in Santa Fe in the years when the Staabs lived there.

Other useful sources on the Staabs and other Jewish families in New Mexico include Henry Tobias's *A History of the Jews in New Mexico;* Floyd Fierman's *Roots and Boots*, a survey of Jewish history in the American Southwest; and Fierman's journal articles, "The Staabs of Santa Fe" and "The Triangle and the Tetragrammaton." William Parish's *The Charles Ilfeld Company* studies the early merchant capitalists in New Mexico. Tomas Jaehn's *Germans in the Southwest, 1850–1920* discusses both New Mexico's German Jews and the general German experience in the desert Southwest.

Other useful histories of New Mexico's territorial years include Paul Horgan's *The Centuries of Santa Fe*, *The Far Southwest* by Howard Lamar, *Blood and Thunder* by Hampton Sides, and the recently published *Chasing the Santa Fe Ring* by David Caffee. Paul Horgan's *Lamy of Santa Fe* is the definitive biography of Archbishop Jean-Baptiste Lamy, offering an intimate and intricate portrait of the world Lamy found and shaped when he arrived in Santa Fe in 1851. Willa Cather's *Death Comes for the Archbishop* looks at the archbishop's life from a fictional vantage point, but her observations of landscape and character in old New Mexico are sharp and crystalline—it is a classic of Southwestern literature. The two most detailed accounts of Abraham Staab's dealings with the archbishop and the cathedral come from Ralph Emerson Twitchell's *Old Santa Fe* and William Keleher's *The Fabulous Frontier*.

In researching Julia Staab's health, I consulted a number of sources on nineteenth-century medicine and the treatment of mental illness in Julia's era. They include: David Dary's *Frontier Medicine*, Judith W. Leavitt's *Sickness and Health in America: Readings in the History of Medicine and Public Health*, Barbara Sicherman's "Uses of a Diagnosis" in the *Journal of the History of Medicine*, Norman Gevitz's *Other Healers*, and Susan Cayleff's *Wash and Be Healed*, a history of the water cure movement in America. Lynn Gamwell and Nancy Tomes's *Madness in America: Cultural and Medical Perceptions of Mental Illness Before 1914* provides a useful overview of the ways nineteenth-century physicians approached mental illness. Sarah Stage's *Female Complaints: Lydia Pinkham and the Business of Women's Medicine* is another fascinating history of women's health and patent medicines in nineteenth-century America.

In researching the nineteenth-century spa movement, I came across a number of detailed and sometimes comical contemporary descriptions of the medicinal benefits of the "healing waters," in Bad Pyrmont, where Julia Staab visited during her 1891 quest for healing, and other German spas. Those books include Thomas Linn's 1894 handbook, *Where to Send Patients Abroad for Mineral and Other Water Cures and Climatic Treatment* and Sigismund Sutro's *Lectures on the German Mineral Waters and on Their Rational Employment for the Cure of Certain Chronic Diseases*. Meanwhile, *Dr Seebohm's Wegweiser in Bad Pyrmont mit Umbegung*, by Adolf Seebohm, lists the specific benefits of the waters of Bad Pyrmont in archaic, flowery, and often laughable Germanic detail.

On the subject of German Jewish history, there is no better book than Amos Elon's *The Pity of It All: A History of the Jews in Germany, 1743–1933*, which begins with Moses Mendelssohn's journey to Berlin and ends with the Nazi takeover. I also consulted W. Michael Blumenthal's *The Invisible Wall: Germans and Jews, a Personal Exploration*, H. G. Adler's *The Jews in Germany: From the Enlightenment to National*

Socialism, and Julia Wood Kramer's *This, Too, Is for the Best: Simon Krämer and His Stories.*

Stephen Birmingham's *"Our Crowd": The Great Jewish Families of New York* is required reading for anyone interested in the German Jewish immigrant experience in New York. Avraham Barkai's *Branching Out: German-Jewish Immigration to the United States, 1820–1914* was also a useful resource. Franz Kafka's *Letter to His Father*, written in 1919, offers anguished insight into the conflicts between German Jewish fathers and their modern sons at the turn of the twentieth century.

For specific details of the Jewish experience in Lügde, I relied on Manfred Willeke's *Genealogie: Die Geschichte der Juden in Lügde (Genealogy: The History of the Jews in Lügde)* and Willy Gerking's article in the *Historisches Handbuch der jüdischen Gemeinschaften in Westfalen und Lippe (Historical Handbook of Jewish Communities in Westphalia and Lippe)*. *The Memoirs of Glückel of Hameln* is essential reading for those seeking to understand the experience of Jewish women in Germany in the late seventeenth and early eighteenth century.

I first learned of Emilie Schuster's death in the Theresienstadt ghetto through my conversations with Wolfgang Mueller, but it was in Margit Naarmaan's history,*"Von Ihren Leuten wohnt hier keiner mehr": Jüdische Familien in Paderborn in der Zeit des Nationalsozialismzus ("None of Your People Live Here Anymore": Jewish Families in Paderborn During the Nazi Time)* that I learned the awful specifics of her loss. Gerhard Schoenberner's *The Yellow Star: The Persecution of the Jews in Europe, 1933–1945* provided documentary information about the deportation of the Paderborn Jews through Bielefeld to Theresienstadt. There are many good histories of Theresienstadt online and in books. I found particularly useful the Terezín museum's publication *Terezín in the "Final Solution of the Jewish Question," 1941–1945: Guide to the Permanent Exhibition of the Ghetto Museum in Terezín*. Jana Renée Friesová's book *Fortress of My Youth: Memoir of a Terezín Survivor*, describes the conditions at Theresienstadt in heartbreaking and horrific detail. Finally, Wolfgang Mueller's memoir, *Wolf: Persecution, Escape, Survival, Triumph* tells of his own personal journey from Nazi Germany to New Mexico and beyond.

On the website Ancestry.com, I located numerous immigration, census, and birth and death records, in addition to a number of family trees that helped me understand my connections to more distant relatives. Other important online genealogical resources I used in tracing my Jewish genealogy and history included the Jewish Virtual Library and the Leo Baeck Institute Archives, where I stumbled upon Ernest Schuster's 1985 "Chronicle of the Schuster Family," which traced the Schusters back to Lügde, Germany, and on through the subsequent generations in Germany and America.

As a neophyte in the world of ghosts and spirits, I had a lot of catching up to do. I am grateful to those historians and science writers who made it easier, first and foremost among them Mary Roach, whose book *Spook: Science Tackles the Afterlife* surveys the world of ghost research and renders it engrossing to the rest of us. Deborah Blum's *Ghost Hunters: William James and the Search for Scientific Proof of Life After Death* is an engaging history of the efforts of the American Society for Psychical Research to cast the light of empirical science on the mediums and psychics who proliferated in the years after the Civil War.

Peter Aykroyd (the comedian Dan Aykroyd's father) wrote a fascinating and comprehensive history of the larger Spiritualism movement in *A History of Ghosts: The True Story of Séances, Mediums, Ghosts, and Ghostbusters*. The author Arthur Conan Doyle, of Sherlock Holmes fame, was also a noted Spiritualist who penned his own chronicle of the movement, *The History of Spiritualism*. Hugh and Susan Harrington published a comprehensive biography, *Annie Abbott: "The Little Georgia Magnet" and the True Story of Dixie Haygood*, of the woman whose levitation act Bertha Staab watched in Los Angeles in 1891, during Spiritualism's heyday.

Judith Richardson's book *Possessions: The History and Uses of Haunting in the Hudson Valley* is a historical study of the supernatural legends of upstate New York that explores how ghost stories—whether we believe them or not—operate as a sort of "social memory" within our cultural landscapes. I also consulted a number of books and websites about the ghosts that inhabit the particular cultural landscape in which I was writing—Santa Fe. Those include Allan Pacheco's *Ghosts-Mayhem-Murder* and Antonio Garcez's *Adobe Angels: The Ghosts of Santa Fe.*

Finally, there were the imaginative works in which Julia played a role: Joanna Hershon's *The German Bride: A Novel* and my third cousin Kay Miller's *Jews of the Wild West: A Multicultural True Story.*

✻ BIBLIOGRAPHY ✻

Interviews and Oral Histories

Baptista, Sarina. Loveland, CO. February 12, 2013.

Blum, Ilene. Boulder, CO. May 31, 2013

Blum, Ilene. Boulder, CO. September 5, 2013.

Conklin, Ed. Cassadaga, FL. March 28, 2013.

Hartman, Betty Mae. Albuquerque, NM. May 14, 2012.

Johnson, Misha. Telephone interview. Boulder CO. February 14, 2013.

Lorenzen, John. Santa Fe, NM. March 29, 2012.

Mangus, Judith. Boulder, CO. December 14, 2012.

Mason, Jonathan. Tesuque, NM. September 20, 2013.

Mueller, Wolfgang. Telephone interview. Boulder CO. May 22, 2012.

Mueller, Wolfgang. Washington, DC. October 15, 2012.

Paxton, Nancy; Nordhaus, Dick and Mary; Messeca, Betsy. Albuquerque, NM. May 14, 2012.

Somers, Juli. Santa Fe, NM. February 19, 2013.

Tobias, Henry. Albuquerque, NM. May 15, 2012.

Wallace, Don. Washington DC. October 16, 2012.

Wallace, Tom. New York, NY. April 24, 2012.

Archives

Floyd Fierman Papers. Bloom Southwest Jewish Archives. University of Arizona.

Staab papers; W. S. Harroun papers. Museum of New Mexico History Archive.

Staab and Spiegelberg papers. New Mexico Jewish Historical Society Archives, New Mexico State Archives.

Television Shows

Ghost Hunters. Season 9, Episode 1. January 16, 2013.

Unsolved Mysteries. Episode 194. October 2, 1994.

City Directories

Los Angeles: 1933

Los Angeles Voter Registry: 1932, 1934, 1940, 1946

New York: 1879, 1916
Oklahoma City: 1905, 1906, 1907, 1908, 1910, 1911, 1912, 1913, 1915

Birth and Death Records, Census Records, Immigration Records

Ancestry.com

jewishvirtuallibrary.org.

genealogybank.com

newspaperarchives.com

newspapers.com

Newspapers

Albuquerque Journal

Albuquerque Morning Democrat

Albuquerque Tribune

Albuquerque Weekly Citizen

American Hebrew

Anaconda Standard

Arizona Weekly Star

Atlanta Constitution

Billings Gazette

Borderer

Boston American

Boston Herald

Boston Record American

Boston Traveler

Colorado Springs Gazette

Daily Oklahoman

Deming Headlight

Denver Post

El Paso Morning Times

El Paso Times

Jewish Spectator

Kansas City Journal

Kansas City Star

Kansas City Times

Las Cruces Daily News

Las Cruces Democrat

Las Vegas Daily Gazette

Las Vegas Optic

Macon Telegraph

New York Herald

New York Herald Tribune

New York Times

Northwest Herald Tribune

Rocky Mountain News

Salt Lake Tribune

San Diego Evening Tribune

Santa Fe Daily Sun

Santa Fe Gazette

Santa Fe Herald

Santa Fe New Mexican

Santa Fe Weekly Gazette

Springfield Republican

Articles

Angel, Frank Warner, and Lee Scott Theisen, ed. "Frank Warner Angel's Notes on New Mexico Territory 1878." *Arizona and the West* 18, no. 4 (Winter 1976): 333–70.

Fierman, Floyd S. "The Staabs of Santa Fe: Pioneer Merchants in New Mexico Territory." *Rio Grande History* no. 13 (1983). Rio Grande Historical Collections, New Mexico State University.

———. "The Triangle and the Tetragrammaton: A Note on the Cathedral at Santa Fe." *New Mexico Historical Review* 37, no. 4 (October, 1962): 310–15.

Gerking, Willy. *Historisches Handbuch der jüdischen Gemeinschaften in Westfalen und Lippe—Die Ortschaften und Territorien im heutigen Regierungsbezirk Detmold.* Ed. Karl Hengst in collaboration with Ursula Olschewski. Publications of the Historical Commission for Westphalia, New Series 10, Münster, Germany (2013): 519–25.

Gilman, Charlotte Perkins. "Why I Wrote 'The Yellow Wallpaper.'" *The Forerunner*, October 1913.

Goldstein, Sherry Gleicher. "Flora Spiegelberg: Grand Lady of the Southwest Frontier." *Southwest Jewish History* 1, no. 2 (Winter 1992).

Old Santa Fe: A Magazine of History, Archaeology, Genealogy and Biography 1, no. 1 (July 1913). Profile of Abraham Staab.

Sicherman, Barbara. "Sickness and Health in America: The Uses of a Diagnosis." *Journal of the History of Medicine and Allied Sciences* 32, no. 1 (January 1977): 33–54.

Spiegelberg, Flora. "Reminiscences of a Jewish Bride of the Santa Fe Trail." *Jewish Spectator*, August and September 1937.

Mostel, Raphael. "Bildung Mendelssohn." *Jewish Daily Forward*, December 25, 2009. http://forward.com/articles/121154/bildung-mendelssohn.

White, Mary Lee. "The Ghost: Julie Staab Still Watches Over Her Home, Now a Part of La Posada." *Santa Fe Reporter*, November 22, 1979.

Books

Ackroyd, Peter, and Angela Narth. *A History of Ghosts: The True Story of Séances, Mediums, Ghosts, and Ghostbusters.* New York: Rodale Books, 2009.

Adler, H. G. *The Jews in Germany: From the Enlightenment to National Socialism.* Notre Dame: University of Notre Dame Press, 1969.

Bandelier, Adolph. *The Southwestern Journals of Adolph F. Bandelier, 1880–1882.* Edited by Charles H. Lange and Carroll L. Riley. Albuquerque: University of New Mexico Press, 1966.

———. *The Southwestern Journals of Adolph F. Bandelier, 1883–1884.* Edited by Charles H. Lange and Carroll L. Riley, with the assistance of Elizabeth M. Lange. Albuquerque: University of New Mexico Press, 1970.

————. *The Southwestern Journals of Adolph F. Bandelier, 1885–1888.* Edited by Charles H. Lange and Carroll L. Riley. Albuquerque: University of New Mexico Press, 1975.

————. *The Southwestern Journals of Adolph F. Bandelier, 1889–1892.* Edited by Charles H. Lange, Carroll L. Riley, and Elizabeth M. Lange. Albuquerque and Santa Fe: University of New Mexico Press, School of American Research, 1984.

Barkai, Avraham. *Branching Out: German-Jewish Immigration to the United States, 1820–1914.* New York: Holmes & Meier, 1994.

Birmingham, Stephen. *"Our Crowd": The Great Jewish Families of New York.* Syracuse: Syracuse University Press, 1967.

Blodig, Vojtěch. *Terezín in the "Final Solution of the Jewish Question," 1941–1945: Guide to the Permanent Exhibition of the Ghetto Museum in Terezín.* Prague: Osvald, 2006.

Blum, Deborah. *Ghost Hunters: William James and the Search for Scientific Proof of Life After Death.* New York: Penguin Books, 2006.

Blumenthal, W. Michael. *The Invisible Wall: Germans and Jews, a Personal Exploration.* Washington, DC: Counterpoint, 1998.

Caffee, David L. *Chasing the Santa Fe Ring: Power and Privilege in Territorial New Mexico.* Albuquerque: University of New Mexico Press, 2014.

Cather, Willa. *Death Comes for the Archbishop.* New York: Alfred A. Knopf, 1927.

Cayleff, Susan. "Gender, Ideology, and the Water-Cure Movement." In *Other Healers: Unorthodox Medicine in America.* Edited by Norman Gevitz, 82–98. Baltimore: Johns Hopkins University Press, 1988.

Cayleff, Susan E. *Wash and Be Healed: The Water-Cure Movement and Women's Health.* Philadelphia: Temple University Press, 1987.

Dary, David. *Frontier Medicine.* New York: Alfred A. Knopf, 2008.

De Waal, Edmund. *The Hare with Amber Eyes.* New York: Farrar, Straus and Giroux, 2010.

Doyle, Arthur Conan. *The History of Spiritualism.* Vol. 1. New York: George H. Doran Company, 1926.

Duden, Gottfried. *Report on a Journey to the Western States of North America and a Stay of Several Years Along the Missouri (During the Years 1824, '25, '26, 1827).* Columbia, MO: University of Missouri Press, 1980.

Eliot, George. *Daniel Deronda.* Public Domain Book.

Elon, Amos. *The Pity of It All: A History of Jews in Germany, 1743–1933.* New York: Henry Holt and Company, 2002.

Fierman, Floyd S. *Roots and Boots: From Crypto-Jew in New Spain to Community Leader in the American Southwest.* Hoboken, NJ: Ktav Publishing House, 1987.

Friesová, Jana Renée. *Fortress of My Youth: Memoir of a Terezín Survivor.* Translated by Elinor Morrisby and Ladislav Rosendorf. Madison: University of Wisconsin Press, 2002.

Gamwell, Lynn, and Nancy Tomes. *Madness in America: Cultural and Medical Perceptions of Mental Illness Before 1914*. Ithaca: Cornell University Press, 1995.

Garcez, Antonio. *Adobe Angels: The Ghosts of Santa Fe*. Santa Fe: Red Rabbit Press, 1992.

Gevitz, Norman, ed. *Other Healers: Unorthodox Medicine in America*. Baltimore: Johns Hopkins University Press, 1988.

Gregg, Josiah. *Commerce of the Prairies; or, The Journal of a Santa Fe Trader during Eight Expeditions*. Vol. 1. Carlisle, MA: Applewood Books, 1851.

Harrington, Susan J., and Hugh T. Harrington. *Annie Abbott: "The Little Georgia Magnet" and the True Story of Dixie Haygood*. Milledgeville, GA: Self-published, 2010.

Hershon, Joanna. *The German Bride: A Novel*. New York: Ballantine Books, 2008.

Horgan, Paul. *The Centuries of Santa Fe*. Boston: E. P. Dutton & Company, 1956.

———. *Lamy of Santa Fe*. Middletown, CT: Wesleyan University Press, 1975.

Jackson, Buzzy. *Shaking the Family Tree: Blue Bloods, Black Sheep, and Other Obsessions of an Accidental Genealogist*. New York: Simon & Schuster, 2010.

Jaehn, Tomas. *Germans in the Southwest, 1850–1920*. Albuquerque: University of New Mexico Press, 2005.

———. *Jewish Pioneers of New Mexico*. Santa Fe: Museum of New Mexico Press, 2004.

Keleher, William. *The Fabulous Frontier: Twelve New Mexico Items*. Albuquerque: University of New Mexico Press, 1962.

Kafka, Franz. *Letter to Father*. Translated by Karen Reppin. Prague: Vitalis/Bibliotheca Bohemica, 2011.

Kramer, Julia Wood. *This, Too, Is for the Best: Simon Krämer and His Stories*. New York: Peter Lang, 1989.

La Farge, John Pen. *Turn Left at the Sleeping Dog: Scripting the Santa Fe Legend, 1920–1955*. Albuquerque: University of New Mexico Press, 2001.

Lamar, Howard. *The Far Southwest: 1846–1912, A Territorial History*. Revised Edition. Albuquerque: University of New Mexico Press, 2000.

Lavender, David Sievert. *The Southwest*. Albuquerque: University of New Mexico Press, 1980.

Leavitt, Judith W. *Sickness and Health in America: Readings in the History of Medicine and Public Health*. Madison, WI: University of Wisconsin Press, 1997.

Linn, Thomas. *Where to Send Patients Abroad for Mineral and Other Water Cures and Climatic Treatment*. Detroit: George S. Davis, 1894.

Lowenthal, Marvin, trans. *The Memoirs of Glückel of Hameln*. New York: Schocken Books, 1977.

Magoffin, Susan Shelby. *Down the Santa Fe Trail and into Mexico: The Diary of Susan Shelby Magoffin, 1846–1847*. Edited by Stella M. Drumm. New Haven, CT: Yale University Press, 1926.

May, Karl Friedrich. *Winnetou, The Apache Knight*. Surry Hills, NSW, Australia: Objective Systems Pty Ltd., 2006.

Melzer, Richard. *Buried Treasures: Famous and Unusual Gravesites in New Mexico History*. Santa Fe: Sunstone Press, 2007.

Meyer, Beatrice Ilfeld. *Don Luis Ilfeld: 1857–1938*. Albuquerque: Albuquerque Historical Society, 1973.

Meyer, Marian. *Mary Donoho: New First Lady of the Santa Fe Trail*. Santa Fe: Ancient City Press, 1991.

Miller, Kay. *Jews of the Wild West: A Multicultural True Story*. Union City, NJ: Paint Horse Press, 2012.

Minces, Elizabeth Nordhaus. *The Family: Early Days in New Mexico*. Albuquerque: Self-published, 1980.

Mueller, Wolfgang. *Wolf: Persecution, Escape, Survival, Triumph*. Bloomington, IN: Abbott Press, 2013.

Naarmann, Margit. *"Von Ihren Leuten wohnt hier keiner mehr": Jüdische Familien in Paderborn in der Zeit des Nationalsozialismzus*. Paderborn, Germany: Paderborner Historische Forschungen Band 7, 1993.

Newman, Albert Harding, and Alfred Johnson. *Harvard Class of 1895—Fifth Report*. Cambridge, MA: Crimson Printing Co., 1915.

Nordhaus, Bertha Staab. *Travel Journals, 1891–1893*. Family collection.

Nordhaus, Jean. *The Porcelain Apes of Moses Mendelssohn*. Boston: Milkweed Editions, 2002.

Nordhaus, Virginia. *Unsent Letters*. Family collection.

Pacheco, Allan. *Ghosts-Mayhem-Murder, Santa Fe Chronicles*. Santa Fe: Sunstone Press, 2004.

Parish, William J. *The Charles Ilfeld Company: A Study of the Rise and Decline of Mercantile Capitalism in New Mexico*. Cambridge, MA: Harvard University Press, 1961.

Richardson, Judith. *Possessions: The History and Uses of Haunting in the Hudson Valley*. Cambridge, MA: Harvard University Press, 2003.

Roach, Mary. *Spook: Science Tackles the Afterlife*. New York: W. W. Norton & Company, 2005.

Segale, Sister Blandina. *At the End of the Santa Fe Trail*. Milwaukee, WI: Bruce Publishing Company, 1948.

Schoenberner, Gerhard. *The Yellow Star: The Persecution of the Jews in Europe, 1933–1945*. New York: Fordham University Press, 2004.

Schuster, Ernest. *Chronicle of the Schuster Family*. Laguna Hills, CA: Ernest Schuster, 1985; Center for Jewish History Digital Collections. http://access .cjh.org/home.php?type=extid&term=1040927#1.

Schwarz, Gary E., with William L. Simon. *The Afterlife Experiments: Breakthrough Scientific Evidence of Life After Death*. New York: Pocket Books, 2002.

Seebohm, Adolf. *Dr Seebohm's Wegweiser in Bad Pyrmont mit Umbegung*. Bad Pyrmont, Germany: Verlag von Ernst Schnelle, Hoflieferant, 1915.

Sides, Hampton. *Blood and Thunder: The Epic Story of Kit Carson and the Conquest of the American West*. New York: Anchor Books, 2006.

Spiegelberg, Flora. *Princess Goldenhair and the Wonderful Flower.* Chicago: Rand McNally & Company, 1915.

Stage, Sarah. *Female Complaints: Lydia Pinkham and the Business of Women's Medicine.* New York: W. W. Norton & Co., 1979.

Stratton, Porter. *The Territorial Press of New Mexico, 1834–1912.* Albuquerque: University of New Mexico Press, 1969.

Sutro, Sigismund. *Lectures on the German Mineral Waters and on Their Rational Employment for the Cure of Certain Chronic Diseases.* London: John W. Parker and Son, 1851.

Tobias, Henry J. *A History of the Jews in New Mexico.* Albuquerque: University of New Mexico Press, 1990.

Todd, R. Larry. *Mendelssohn: A Life in Music.* New York: Oxford University Press, 2003.

Twitchell, Ralph Emerson. *Old Santa Fe.* Santa Fe: Sunstone Press, 2007.

Willeke, Manfred, *Genealogie: Die Geschichte der Juden in Lügde.* Lügde: Self-published, 1990.

———. *Lügde: Die Reihe Archivbilder.* Erfurt: Alan Sutton, 2000.

———. *Lügder Sagen-Sammlung: Sagen und sagenhafte Geschichten aus der Stadt Lügde.* Lügde: Self-published, 1988.

❋ PHOTOGRAPHY CREDITS ❋

Page 1: Paul Horgan, from *The Centuries of Santa Fe*, 1956, Courtesy of Special Collections and Archives, Wesleyan University.

Page 18: Family collection.

Page 34: From *Santa Fe Railroad: By the Way* (Chicago: Rand McNally and Company, 1922).

Page 41: C. G. Kaadt, Courtesy of Palace of the Governors Photo Archives (NMHM/DCA), Neg. No. 11070, circa 1895.

Page 47: Courtesy of Palace of the Governors Photo Archives, (NMHM/DCA), Neg. No. 7890.

Page 59: Family collection.

Page 64: Courtesy of Palace of the Governors Photo Archives (NMHM/DCA), Neg. No. 67735.

Page 77: William Henry Brown, Courtesy of Palace of the Governors Photo Archives (NMHM/DCA), Neg. No. 9970, circa 1900.

Page 88: Family collection, 1937.

Page 98: Reverend John C. Gullette, Courtesy Palace of the Governors Photo Archives (NMHM/DCA), Neg. No. 13250, circa 1870.

Page 111: Family collection.

Page 123: Courtesy of the author.

Page 131: Courtesy of Susan J. Harrington.

Page 149: Courtesy of Museum of New Mexico Press, circa 1895.

Page 154: From Joseph Buchanan, *Outlines of Lectures on the Neurological System of Anthropology* (Cincinnati: Journal of Man, 1854).

Page 164: Dana Chase, Courtesy of Palace of the Governors Photo Archives (NMHM/DCA), Neg. No. 56980.

Page 175: Courtesy of the author.

Page 182: Family collection.

Page 191: Courtesy of Felix Warburg.

Page 198: Harmon Parkhurst, Courtesy Palace of the Governors Photo Archives (NMHM/DCA), Neg. No. 10779, 1925.

Page 205: Family collection.

Page 212: Courtesy of Sonya Mueller.

Page 220: Courtesy of Margit Naarmann.

Page 227: PT 1693, Terezín Memorial, courtesy of Tomáš Fritta-Haas.

Page 242: Family collection.

Page 249: Courtesy of Palace of the Governors Photo Archives (NMHM/DCA), Neg. No. 11040.

Page 257: Family collection.

Page 266: Family collection.

Page 272: Courtesy of La Posada de Santa Fe.

Page 285: Courtesy of the author.

Page 291: Courtesy of the author.

About the author

About the book

Insights,
Interviews
& More . . .

Read on

Meet Hannah Nordhaus

Casie Zalud

HANNAH NORDHAUS is the author of the critically acclaimed national bestseller *The Beekeeper's Lament*, which was a PEN Center USA Book Awards finalist, Colorado Book Awards finalist, and National Federation of Press Women Book Award winner. She has written for the *Financial Times*, *Los Angeles Times*, *Outside* magazine, *Times Literary Supplement*, *Village Voice*, and many other publications. She lives with her husband and two children in Boulder, Colorado. ∾

Q&A with Hannah Nordhaus

You write in the book that you've known about your great-great-grandmother Julia Staab your entire life. Why did you decide to write about her now?

I had always found Julia interesting, of course, and I had even written about her when I was a young woman starting out as a journalist. But she was always more of an anecdote to me than a real person with a real story.

Shortly after I gave birth to my first child, however, I was poking around the dusty bookshelves in the house my great-grandfather built in the mountains east of Santa Fe and found a family history my great-aunt Lizzie had written shortly before she died in 1980. I may have read it without much interest when I was a young girl. But this time around, I found it riveting: Lizzie told a tale of sadness and madness and forbidden love, of drug addictions and suicides, knives to the "bosom," inheritance and disinheritance, penury, family feuds, brother against brother. There was, I realized, more to Julia's story than just a ghost in an old hotel. Julia had actually been alive once; she had been dragged as a new bride across the Santa Fe Trail to an unfamiliar place; she had been a new mother, like me. I now wanted to learn more about her.

I also realized that if I wanted to go deeper into my family's history, it had to happen quickly. None of my living ▶

relatives remembered Julia, who died in 1896, but some of them did remember her children (the last one died in 1968), and it seemed important to capture those recollections before they were lost.

By then, too, the Internet had begun to revolutionize genealogy and history research—so many documents are now scanned and searchable online that were, only five or ten years ago, tucked away in dusty archives and difficult to locate. This was the time to do it.

Do you see your book as a ghost story, a biography, or something else?

I think of this book as a history that is wrapped in a ghost story. The story—the legend—of Julia's life, death, and afterlife is what makes her interesting to most people and what keeps her alive to us so many years after her death. It is the reason people want to know about Julia in the first place.

But it's also very much a device through which I was able to explore the other, equally intriguing stories of my family's past, and all the different pasts that bear on Julia's story—Jewish history in eighteenth- and nineteenth-century Germany; the settling of the Anglo-American Southwest; the European spas and séance rooms of the late nineteenth century; the fate of German Jews during the World War II era; and even the eighties and nineties, when Julia's ghost story first entered our cultural imagination.

The ghost story is like an open window (cue spooky music): it beckons us to look through it and rewards us with a view of the past.

There are so many different lenses through which you look to understand who Julia Staab was: the spiritualist movement, Jewish history, the history of the American Southwest, etc. What context did you find gave the most insight into the mystery of Julia's life?

These different strands of history were all essential to telling Julia's story; I've braided them together to create a larger portrait of Julia's life and time. But I'd say that the most helpful to my understanding was the specific history of the Southwest that I encountered in old travel journals and newspapers from the time: what it was like to travel the Santa Fe Trail in the early days, and how it must have been for Julia as a young bride riding into Santa Fe, then a ragged frontier town of trash-strewn dusty streets and brothels and gunfights and all-night fandangos.

In addition, the history of nineteenth-century medicine—particularly its treatment of women's health and mental health—was really invaluable in helping me understand Julia's physical and emotional condition, and the strange and barbaric treatments she would have undergone in hopes of recovery.

You visit multiple psychics and spiritualists in an attempt to connect with Julia's ghost. Did you find this helpful, and why or why not?

My visits to the spiritualists were lots of fun, but they also added a quotient of real human feeling to my search.

Often, I was able to trace Julia's story only through the stories of others—her husband, Abraham, her children, and other people whose better-documented lives intersected hers. What the psychics did was to provide me with a means of connecting to Julia herself—if not to her actual spirit, then at the very least to an idea of her. They told me that she liked flowers, that she had once loved another man, that her children were the world to her, that she rocked back and forth in a rocking chair and brushed her white hair and paced the floor and wrung her hands. To them, she was a woman with a story of her own. For all the unreality of sitting in a room talking to strangers who were themselves talking to "spirits" who resided in the air, my meetings with the psychics made Julia feel somehow more real to me.

Toward the end of the book, you leap forward in time to tell the story of Julia's sister Emilie, who perished in the Holocaust. Why did you decide to include that story in the book?

When I started researching the book, I hadn't planned to extend my narrative far beyond Julia's death. I did, however, hope that ▶

I would learn what had happened to those relatives of Julia's who had remained in Germany.

Then I met my grandfather's second cousin Wolfgang Mueller, who had come to stay with my family in New Mexico as a German Jewish refugee in 1936. He told me that his grandmother, Julia's youngest sister, Emilie, lived long enough to die in a Nazi concentration camp, and I realized that this was a story I needed to explore. Not only because the specter of the Holocaust haunts every German Jewish story, regardless of when and where it ends, but also because it seemed as if Emilie's life in Germany— cultured, privileged, surrounded by friends and family—was everything that Julia felt she had lost in coming to America.

In Mary Doria Russell's wonderful novel *Doc*—a fictional rendering of another legend of the Old West—Doc Holliday cautions his friend Wyatt Earp against remaining mired in a "ghost life"—the life one might have had, if things had gone differently. Emilie was, in many ways, leading the very ghost life that Julia had longed for in her New Mexican exile. But it ended in an unimaginably horrible manner. Emilie's story, writ large, haunts so many of us. Hers was an alternate ending to Julia's life and immigration story, and I felt compelled to explore it.

What is one thing you hope readers will take away from American Ghost?

I hope that readers finish *American Ghost* feeling as connected to the past as I did while researching and writing it. I spent months scrolling through old newspapers, reading travel journals, and paging through diaries, and there were times that I felt so close to Julia's world that I could almost touch it. It was as if I were there with her and her daughters, wearing a high-necked Victorian dress and collecting silver spoons.

I hope that the book can bring that sense of Julia's world to readers, and help them understand, as I wrote of Julia and Abraham's house, that "every rosette and fixture and plaster molding had a dream behind it, a story that began in rural Germany and ended in the hands of strangers." Almost every old building, every gravestone, every street contains a history that spans generations and continents. This is true in Europe, of course, but it is also true in the American West. The "Old West" was not just a place of cowboys and Indians and gunfights and player pianos that sprang up in the desert out of nothing; it was indissolubly tied to the markets in the eastern United States and in Europe, and also to the lives of the many people who settled or passed through from elsewhere. The drama of Julia's life may have unfolded on the isolated frontier, but it's also a larger story of immigration, of assimilation, of religion, of capitalism, of gender, of expectations, of how legends are formed. The past can seem so far away, but it's right there for us to see, if only we look. ᴄᴧ

Unforgiving Lands
A Young Bride Among the Roustabouts of Santa Fe

WHEN MY GREAT-GREAT-GRANDMOTHER set out for New Mexico Territory in 1866, she spoke no English. Nor did she speak any Spanish. German was her native language, Yiddish as well. Julia Staab was a German Jew from a small village in Prussia. I don't know how her marriage to my great-great-grandfather Abraham Staab came about—if it was arranged beforehand, or if they chose each other. But I do know that they were in a hurry to begin their married life in Santa Fe—to inhabit their American Dream.

Abraham was, anyway. He had left their village a decade earlier, at fifteen, to make his fortune. That he did, hauling merchandise—"Hats Boots & Shoes, Hardware, Groceries etc. etc."—along the Santa Fe Trail between St. Louis and the American Southwest. He became a US citizen on July 10, 1865, only a few weeks after the last shots of the Civil War were fired, and promptly departed for Germany in search of a bride. My great-great-grandparents were married on Christmas Day 1865. Julia was twenty-one years old, Abraham twenty-six.

They shipped out on the RMS *Scotia*, a luxury liner that was at the time the fastest ship on the Atlantic, and on January 12 they landed in New York.

From there, they climbed onto a train, and then a steamboat, and then rode for two weeks in a stagecoach across the snow-cloaked Great Plains to make a life among New Mexico's stark and rugged Sangre de Cristo Mountains.

Santa Fe, in 1866, was not yet the elegant city of artists and tourists and well-heeled retirees. It was a rough and unruly town, sandy and treeless. Its central plaza was crowded with carts, wagons, teamsters, roustabouts, soldiers, veterans, fortune seekers, consumptives, Navajos, Apaches, Jewish merchants, freed slaves, miners, gamblers, prostitutes, shysters, horses, burros, pigs, and goats—a confusion of commerce, a babel of languages. The houses were constructed of mud, the streets clouded with billowing dust. Beyond the town's edges stretched a bewildering landscape of uncompromising sky and chisel-topped *cerritos*, so different from anything a young bride from the green and gentle valleys of northwestern Germany would ever have seen. New Mexico was all tans and reds, the ground littered with rocks and reptiles, with hematite-seeped rocks and bleached bones and spiny flora—cactus, greasewood, Spanish bayonet.

This desert was, certainly, an unforgiving land. But it was nonetheless a place that seemed willing to forgive the fact that Julia and Abraham were Jews. In Lügde, the village in which they were raised, local records describe an 1866 cholera outbreak that killed "126 people and one Jew." That Jew was Julia's cousin Philipp Schuster—singled out because, in Julia's time, a Jew in Lügde was not a person but an invasive species, taxed and fined and snubbed at every turn.

Not so in Santa Fe. In the New Mexico that Julia encountered in 1866, the newspapers of the territory spoke kindly of the local Jews ("Many of the best residents are of the Jewish faith," wrote the *Santa Fe New Mexican*). Perhaps this was because the Jewish merchants were advertisers, or perhaps because there weren't enough of them to seem threatening. There were, in Santa Fe, no temples, no Hebrew schools, no Jewish ghettos. The stores stayed open on Saturdays; a rabbi traveled from Denver every few years to circumcise the boys. My great-grandmother Bertha's diaries from those days mention riding parties and sewing circles and teas and Christmas celebrations with gentile and Jewish friends alike—champagne and oysters, boxes at the Albuquerque ▶

opera. But not once in the diary did she mention the fact that her family was Jewish. It didn't seem to matter.

The Staabs were American. They occupied the heart of Santa Fe, with a huge storefront right on the Plaza and a towering family mansion—a mansard-roofed French Second Empire–style brick building—just a few blocks away. The three Staab girls rode sidesaddle and carried gold-headed riding crops. The four boys wore tennis whites and striped sweaters. Abraham was elected county commissioner twice; he helped bring the railroad, the gasworks, and the territorial prison to Santa Fe. He prospered alongside this former Mexican outpost. Brick by brick, railroad tie by railroad tie, he worked to transform Santa Fe from a foreign colony into an American city. The town was parched and unkempt and far from the "civilized" world. But Abraham flourished in that hard soil.

Julia did not. She struggled there; indeed, she seemed to wither in the desert. She bore seven children in quick succession, and lost an eighth. She suffered miscarriages and health problems, and from "hysteria," as they called it then. Whenever she fell into a decline, she traveled to Germany to recover, visiting health spas and German doctors and her many sisters who lived there and tended her when she was unwell. Julia was the only one of the family's eight girls to leave Germany. She felt terribly unlucky to have done so.

In her last years, Julia shut herself in and never left the upstairs bedroom of the European brick home her husband had built among the adobes. While the family celebrated weddings on the ground floor, she stayed upstairs in her room, and she died there in 1896. It is said that her ghost still haunts the building. And that she was also haunted—by the life she might have lived in Germany, and all that she had left behind.

Of course we, who came after, know what became of all that she left behind—what became of her nieces and nephews and of her sister Emilie, who lived long enough to die, at the age of eighty-one in a Nazi concentration camp. We know how it ended. And we are haunted by a ghost life too—the life that might have

been ours had Abraham not dragged Julia across the ocean and plains to this open desert land.

To become American is to accept a staggering loss of self—of the people we once were, in the places we once came from. It may take a generation, perhaps two. But inevitably, it transpires. The surge of conquering culture sweeps down through the generations, much as the spring floods scour the desert arroyos. Washed away, we must lay down new roots.

Julia believed her life in the desert was a curse. But five generations downstream, I find that I can't agree with her. That sere and serrated Western landscape is the only place I have ever felt at home. My father and grandfather came from there, my great-grandmother too. The high desert is in my blood. And I can only see that it was a blessing. ∾

Originally written for What It Means to Be American, a national conversation hosted by the Smithsonian (http://americanhistory.si.edu/) and Zócalo Public Square (http://www.zocalopublic square.org/2015/06/16/a-young-bride-among-the-roustabouts-of-santa-fe/chronicles/who-we-were/). Reproduced with permission.